THE RECURSIVE UNIVERSE

THE
RECURSIVE
UNIVERSE

Cosmic Complexity
and the Limits of
Scientific Knowledge

WILLIAM POUNDSTONE

Computer Consultation
by Robert T. Wainwright

WILLIAM MORROW AND COMPANY, INC. | NEW YORK

Library of Congress Cataloging in Publication Data

Poundstone, William.
The recursive universe.

Bibliography: p.
Includes index.
1. Self-organizing systems. 2. Machine theory.
I. Wainwright, Robert T. II. Title.
Q325.P68   1984   003   84-9045
ISBN 0-688-03975-8

Printed in the United States of America

First Edition

1 2 3 4 5 6 7 8 9 10

BOOK DESIGN BY VICTORIA HARTMAN

To my parents

ACKNOWLEDGMENTS

Martin Gardner introduced Life through his October 1970 *Scientific American* column. Interest in the game soon spawned a newsletter, *Lifeline,* published by Robert T. Wainwright from 1971 to 1973. Most of what is known of the Life universe is the work of *Lifeline*'s many correspondents. Among those who have contributed to this book through their discoveries and insights are Simon Norton, Gary Filipski, Brad Morgan, Ranan B. Banerji, D. M. Saul, Robert April, Michael Beeler, R. William Gosper, Jr., Richard Howell, Rich Schroeppel, Michael Speciner, Keith McClelland, Thomas Holmes, Michael Sporer, Philip Stanley, Donald Woods, William Woods, Roger Banks, Sol Goodman, Arthur C. Taber, Robert Bison, David W. Bray, Charles L. Corderman, Gary Goodman, Stephen B. Gray, Maxwell E. Manowski, Clement A. Lessner III, William P. Webb, Hugh Thompson, Robert Kraus, Rici Liknaitzky, Bill Mann, Steve Ward, James F. Harford, Curt Gibson, Jan Kok, Douglas G. Petrie, Philip Cohen, Paul Wilson, V. Everett Boyer, Dave Buckingham, Mark Niemiec, Peter Raynham, Dean Hickerson, Paul Schick, and George D. Collins, Jr. The information on relative frequencies of Life objects is largely the work of Hugh Thompson. The photograph of the breeder is from Gosper's group at MIT.

CONTENTS

Acknowledgments 7

1 · Complexity and Simplicity 13

2 · The Life Universe 33

3 · Maxwell's Demon 52

4 · Gliders and Spaceships 78

5 · Information and Structure 90

6 · Unlimited Growth 103

7 · Physics as Recursion 122

8 · Recursive Games 133

9 · Big Bang and Heat Death 142

10 · Random Fields 164

11 · Von Neumann and
Self-Reproducing Machines 177

12 · Self-Reproducing Life Patterns 196

13 · The Recursive Universe 230

Life for Home Computers 233

Bibliography 243

Index 247

THE RECURSIVE UNIVERSE

COMPLEXITY AND SIMPLICITY

In the early 1950s, the Hungarian-American mathematician John Von Neumann was toying with the idea of machines that make machines. The manufacture of cars and electrical appliances was becoming increasingly automated. It wasn't hard to foresee a day when these products would roll off assembly lines with no human intervention whatsoever. What interested Von Neumann particularly was the notion of a machine that could manufacture *itself*. It would be a robot, actually. It might wander around a warehouse, collecting all the components needed to make a copy of itself. When it had everything, it would assemble the parts into a new machine. Then both machines would duplicate and there would be four . . . and then eight . . . and then sixteen . . .

Von Neumann wasn't interested in making a race of monster machines; he just wondered if such a thing was possible. Or does the idea of a machine that manufactures itself involve some logical contradiction?

Von Neumann further wondered if a machine could make a machine more complex than itself. Then the machine's descendants would grow ever more elaborate, with no limit to their complexity.

These issues fascinated Von Neumann because they were so fundamental. They were the sorts of questions that any bright child might ask, and yet no mathematician, philosopher, scientist, or engineer of the time could answer them. About all that anyone could say was that all existing machines manufactured machines much simpler than themselves. A labyrinthine factory might make a can opener.

Many of Von Neumann's contemporaries were interested in automatons as well. Several postwar college campuses boasted a professor who had built a wry sort of robot pet in a vacant lab or garage. There was Claude Shannon's "Theseus," a maze-solving electric rodent; Ross Ashby's "machina spora," an automated "fireside cat or dog which only stirs when disturbed," by one account; and W. Grey Walter's restless "tortoise." The tortoise scooted about on motorized wheels, reacting to obstacles in its path but tending to become fouled in carpet shag. When its power ran low, the tortoise refueled from special outlets.

Von Neumann's hunch was that self-reproducing machines are possible. But he suspected that it would be impractical to build one with 1950s technology. He felt that self-reproducing machines must meet or exceed a certain minimum level of complexity. This level of complexity would be difficult to implement with vacuum tubes, relays, and like components. Further, a self-reproducing automaton would have to be a full-fledged robot. It would have to "see" well enough to recognize needed components. It would require a mechanical arm supple enough to grip vacuum tubes without crushing or dropping them, agile enough to work a screwdriver or soldering iron. As much as Von Neumann felt a machine could handle these tasks in principle, it was clear that he would never live to see it.

Inspiration came from an unlikely source. Von Neumann had supervised the design of the computers used for the Manhattan Project. For the Los Alamos scientists, the computers were a novelty. Many played with the computers after hours.

Mathematician Stanislaw M. Ulam liked to invent pattern games for the computers. Given certain fixed rules, the computer would print out ever-changing patterns. Many patterns grew almost as if they were alive. A simple square would evolve into a delicate, corallike growth. Two patterns would "fight" over territory, sometimes leading to mutual annihilation. Ulam developed three-dimensional games too, constructing thickets of colored cubes as prescribed by computer.

Ulam called the patterns "recursively defined geometric objects." *Recursive* is a mathematical term for a repeated procedure, in this case, the repeated rules by which the computers generated the patterns. Ulam found the growth of patterns to defy analysis. The patterns seem to exist in an abstract world with its own physical laws.

Ulam suggested that Von Neumann "construct" an abstract universe for his analysis of machine reproduction. It would be an imaginary world with self-consistent rules, as in Ulam's computer games. It would be a world complex enough to embrace all the essentials of machine operation but otherwise as simple as possible. The rules governing the world would be a simplified physics. A proof of machine reproduction ought to be easier to devise in such an imaginary world, as all the nonessential points of engineering would be stripped away.

The idea appealed to Von Neumann. He was used to thinking of computers and other machines in terms of circuit or logic diagrams. A circuit diagram is a two-dimensional drawing, yet it can represent any conceivable three-dimensional electronic device. Von Neumann therefore made his imaginary world two-dimensional.

Ulam's games were "cellular" games. Each pattern was composed of square (or sometimes triangular or hexagonal) cells. In effect, the games were played on limitless checkerboards. All growth and change of patterns took place in discrete jumps. From moment to moment, the fate of a given cell depended only on the states of its neighboring cells.

The advantage to the cellular structure is that it allows a much simpler "physics." Basically, Von Neumann wanted to create a world of animated logic diagrams. Without the cellular structure, there would be infinitely many possible connections between components. The rules needed to govern the abstract world would probably be complicated.

So Von Neumann adopted an infinite checkerboard as his universe. Each square cell could be in any of a number of states corresponding roughly to machine components. A "machine" was a pattern of such cells.

Von Neumann could have allowed a distinct cellular state for every possible component of a machine. The fewer the states, the simpler the physics, however. After some juggling, he settled on a cellular array with 29 different states for its cells. Twenty-eight of the states are simple machine components; one is the empty state of unoccupied cells. The state of a cell in the next moment of time depends only on its current state and the states of its four bordering ("orthogonal") neighbors.

Von Neumann's cellular space can be thought of as an exotic, solitaire form of chess. The board is limitless, and each square can be empty or contain one of 28 types of game pieces. The lone player arranges game pieces in an initial pattern. From then on, strict rules determine all successive configurations of the board.

Since the player has no further say, one might as well imagine that the game's rules are automatically carried out from one move to the next. Then the player need only sit back and watch the changing patterns of game pieces that evolve.

What Von Neumann did is this: He proved that there are starting patterns that can reproduce themselves. Start with a self-reproducing pattern, let the rules of the cellular space take their course, and there will eventually be two patterns, and then four, and then eight . . .

Von Neumann's pattern, or machine, reproduced in a very general, powerful way. It contained a complete description of its own organization. It used that information to build a new copy of itself. Von Neumann's machine reproduction was more akin to the reproduction of living organisms than to the growth of crystals, for instance. Von Neumann's suspicion that a self-reproducing machine would have to be complicated was right. Even in his simplified checkerboard universe, a self-reproducing pattern required about 200,000 squares.

Since Von Neumann was able to prove that a self-reproducing machine is possible in an imaginary but logical world, no logical contradiction must be inherent in the concept of a self-reproducing machine. Ergo, a self-reproducing machine is possible in our world. No one has yet made a self-reproducing machine, but today no logician—or engineer—doubts that it is possible.

VON NEUMANN'S MACHINE AND BIOLOGY

Not only can a machine manufacture itself, but Von Neumann was also able to show that a machine can build a machine more complex than itself. As it happens, these facts have been of almost no use (thus far, at least) to the designers of machinery. Von Neumann's hypothetical automatons have probably had their greatest impact in biology.

One of the longest running controversies in biology is whether the actions of living organisms can be reduced to chemistry and physics. Living matter is so different from non-living matter that it seems something more than mere chemistry and physics must be at work. On the other hand, perhaps the organization of living matter is so intricate that the observed properties of living organisms do follow, ultimately, from chemistry and physics.

The latter viewpoint is held by virtually all biologists today, but it is not particularly new. René Descartes believed the human body to be a machine. By that he meant that the body is composed of substances interacting in predictable ways according to the same physical laws that apply to nonliving matter. Descartes felt the body is understandable and predictable, just as a mechanical device is.

Queen Christina of Sweden, a pupil of Descartes, questioned this belief: How can a machine reproduce itself? No known inanimate object reproduced in the way that living organisms do.

Not until Von Neumann did anyone have a good rebuttal to Christina's objection. Many biologists postulated a "life force" to explain reproduction and other properties of living matter. The life force was the reason that living matter is so different from nonliving matter. It supplemented the ordinary physical laws in living matter. Because the life force could never apply to a mechanical device, it was pointless to compare living organisms to machines.

At the simplest level, Von Neumann's self-reproducing machine is nothing like a living organism. It contains no DNA, no protein; it is not even made of atoms. To Von Neumann such considerations were beside the point. Von Neumann's abstract machine demonstrates that self-reproduction is possible solely within the context of physics and logic. No *ad hoc* life force is necessary.

It became reasonable to see living cells as complex self-reproducing machines. A living cell must perform many tasks besides reproducing itself, of course. So it is not surprising that real cells are much more difficult to understand than Von Neumann's machine.

By coincidence, Watson and Crick's work on DNA synthesis took place concurrently with much of Von Neumann's study of machine reproduction. By about 1951, molecular biologists had identified structural proteins, enzymes, nucleotides, and most other important components of cells. The genetic information of cells was known to be encoded in observable structures (chromosomes) composed of nucleotides and certain proteins. Still, no one had any concrete idea how these components managed to reproduce themselves. It might require some sort of a life force yet.

The discovery of DNA's structure and the genetic code were first steps in understanding how assemblages of molecules implement their own reproduction. It is remarkable that the logical basis of reproduction in living cells is almost identical to that of Von Neumann's machine. There is a part of Von Neumann's machine that corresponds to DNA, another part that corresponds to ribosomes, another part that does the job of certain enzymes. All this tends to confirm that Von Neumann was successful in formulating the simplest truly general form of self-reproduction.

Biologists have adopted Von Neumann's view that the essence of self-reproduction is organization—the ability of a system to contain a complete description of itself and use that information to create new copies. The life force idea was dropped. Biology is now seen as a derivative science whose laws can (in principle) be derived from more fundamental laws of chemistry and physics.

IS PHYSICS DERIVATIVE TOO?

"What really interests me," Albert Einstein once remarked, "is whether God had any choice in the creation of the world." Einstein was wondering if the world and its physical laws are arbitrary or if they are somehow inevitable. To put it another way: Could the universe possibly be any different than it is?

In many ways, the world seems arbitrary. An electron

weighs 0.000000000000000000000000000091096 grams. The stars cluster in galaxies, some spiral and some elliptical. No one knows of any obvious reason why an electron couldn't be a little heavier or why stars aren't spaced evenly through space. It is not hard to imagine a world in which a few physical constants are different or a world in which *everything* is different.

Einstein's statement draws a variety of reactions. One's feelings about the nature or reality of God are largely immaterial. The issue is whether there can be "prior" constraints even on the creation of the world.

One reaction is that there are no such constraints. By definition, an all-powerful God can create the world any way He chooses. Everything about the universe is God's choice. Creation was not limited, not even by logic. God created logic.

On the other hand, it can be argued that even God is bound by logic. If God can lift any weight, then he is expressly prevented from being able to create a weight so heavy that He cannot lift it. But God can do anything that does not involve a logical contradiction.

Einstein evidently sympathized with the latter argument. He realized that it may not be easy to recognize logical contradictions, however. Suppose that by some recondite chain of reasoning it is possible to show that the speed of light *must* be exactly 299,792.456 kilometers per second. Then the notion of a world in which the speed of light is even slightly different would involve a contradiction.

Conceivably, all the laws of physics could have a justification in pure logic. Then God would have had no freedom whatsoever in determining physics.

Taking this idea further, perhaps the initial conditions of our world were also determined by arcane logic. The big bang had to occur precisely the way it did; the galaxies had to form just the way they did.

In that case, God would have had no choice in creation at all. He could not even have made a world in which forks were placed on the right side of plates or in which New Zealand didn't exist. This would be the only possible world.

Some people dismiss Einstein's statement as too metaphysical—the sort of thing that could be argued endlessly without getting anywhere. In fact, modern physics has much to say about choice in creation.

More and more physical laws have turned out to be derivative. If law B is the inevitable consequence of law A, then only law A is really an independent law of nature. B is a mere logical implication.

The chain of implication can have many links. A biological phenomenon may be explained in terms of chemistry. The chemistry may be explained in terms of atomic physics. The atomic physics may be explained in terms of subatomic physics. Then the laws of subatomic physics suffice to explain the original biology.

This is what the reductionist mode of thought seeks. To understand a phenomenon is to be able to give reasons for it —to see the phenomenon as inevitable rather than arbitrary. Reductionism has been a keystone of Western science. Einstein wondered if reductionism can proceed endlessly or whether it will reach a dead end.

Marquis Pierre Simon de Laplace, one of the founders of the reductionist tradition, believed that everything is knowable. He illustrated his point with a hypothetical cosmic intelligence that might, in principle, know the details of every atom in the cosmos. Given a knowledge of the laws of physics, the being should be able to predict the future and reconstruct the past, Laplace felt: "Nothing would be uncertain and the future, as the past, could be present to its eyes."

Physicists are close to a comprehensive theory of nature in which all the world's phenomena are reduced to just two fundamentally different types of particles interacting by just two types of force. This long-awaited synthesis goes by the name of grand unified theory (GUT). Some physicists contemplate an even broader theory in which everything is reduced to one particle and one force. This is labeled a super-grand unified theory. It is always dangerous to suppose that physicists have or have nearly learned everything there is to know about physics. Laplace's cosmic being was evidently prompted by his belief that eighteenth-century physics was nearly complete; late nineteenth-century physicists expressed similar beliefs just in time to have them demolished by quantum theory and relativity. Nonetheless, physicists are (again) facing the prospect of a comprehensive theory of nature, one that may lie in the completion of current theoretical projects. It is a good time to reexamine the cornerstones of reductionism.

This book cannot cover the grand and super-grand unified theories except tangentially. Nor does it mean to endorse such theories as the last word in natural law. Its purpose is rather to look at the explaining power of physical law. Not since the turn of the century have physicists believed physics to be substantially complete. The new field of information theory has grown up in the interim. Information theory will have extensive philosophical consequences for any future comprehensive theory of nature.

THE LIMITS OF KNOWLEDGE

Laplace, Leibniz, Descartes, and Kant popularized the idea of the universe as a vast machine. The cosmos was likened to a watch, composed of many parts interacting in predictable ways. Implicit in this analogy was the possibility of predicting all phenomena. The perfect knowledge of Laplace's cosmic intelligence might be a fiction, but it could be approximated as closely as desired.

Today the most common reaction to Laplace's idea, among people who have some feel for modern physics, is to cite quantum uncertainty as its downfall. In the early twentieth century, physicists discovered that nature is nondeterministic at the subatomic scale. Chance enters into any description of quanta, and all attempts to exorcise it have failed. Since there is no way of predicting the behavior of an electron with certainty, there is no way of predicting the fate of the universe.

The quantum objection is valid, but it is not the most fundamental one. In 1929, two years after Heisenberg formulated the principle of quantum uncertainty, Hungarian physicist Leo Szilard discovered a far more general limitation on empirical knowledge. Szilard's limitations would apply even in a world not subject to quantum uncertainty. In a sense, they go beyond physics and derive instead from the logical premise of observation. For twenty years, Szilard's work was ignored or misunderstood. Then in the late 1940s and early 1950s it became appreciated as a forerunner of the information theory devised by Claude Shannon, John Von Neumann, and Warren Weaver.

Information theory shows that Laplace's perfect knowledge is a mirage that will forever recede into the distance. Science is empirical. It is based solely on observations. But observation is a two-edged sword. Information theory claims that every observation obscures at least as much information as it reveals. No observation makes an information profit. Therefore, no amount of observation will ever reveal everything— or even take us any closer to knowing everything.

Complementing this austere side of information theory is insight into how physical laws generate phenomena. Physicists have increasing reason to suppose that the phenomena of the world, from galaxies to human life, are dictated more by the fundamental laws of physics than by some special initial state of the world. In terms of Einstein's riddle, this suggests that God had little meaningful choice once the laws of physics were selected. Information theory is particularly useful in explaining the complexity of the world.

THE PARADOX OF COMPLEXITY

If the grand unified theories are the triumph of reductionism, they also make it seem all the more preposterous. The GUTs postulate two types of particles, two types of forces, and a completely homogeneous early universe. How could any structure arise from such a simple initial state?

The world's staggering diversity seems to preclude any simple explanation. There is structure at all scales, from protons to clusters of galaxies. Even before unified theories of physics, the world's richness seemed a paradox.

It is easy to show that the world is far, far more complex than can be accounted for by simple interpretations of chance. Take, for instance, the old fantasy of a monkey typing *Hamlet* by accident. If there are 50 keys on a typewriter, then the chance of the monkey hitting the right key at any given point is 1 in 50. There are approximately 150,000 letters, spaces, and punctuation marks in a typical text of *Hamlet*. Once the monkey has struck the keyboard 150,000 times, the chance that it has produced *Hamlet* is 1 in 50 multiplied by itself 150,000 times.

Fifty multiplied by itself 150,000 times (which can be written $50^{150,000}$) is an unimaginably huge number. It cannot even be called an astronomical number, for it is much larger than any number with astronomical significance. Just writing $50^{150,000}$ out would take about 255,000 digits and fill about half this book.

In contrast, all of the large numbers encountered in physics can be written out easily. It is estimated that the number of fundamental particles in the observable universe is (give or take a few zeros) 1,000,000,000,000,000,000,000,000,000,-000,000,000,000,000,000,000,000,000,000,000,000,-000,000,000,000. Vast as this number is, it is nothing compared to $50^{150,000}$.

In view of this, it may seem remarkable that anything as complex as a text of *Hamlet* exists. The observation that *Hamlet* was written by Shakespeare and not some random agency only transfers the problem. Shakespeare, like everything else in the world, must have arisen (ultimately) from a homogeneous early universe. Any way you look at it, *Hamlet* is a product of that primeval chaos.

If every particle in the universe were replaced with a monkey and a typewriter and all the monkeys had been striking keys since the big bang, the chance of producing *Hamlet* would still be negligible. Yet *Hamlet* was produced from a series of physical processes that (initially, at least) were even more chaotic than monkeys banging at typewriters.

Of course, *Hamlet* is just one example of the complexity of the world. The world is full of things that are too unlikely to be mere accidents. If the GUTs are to be comprehensive, then *Hamlet*, Shakespeare, the earth, and galaxies must all be consequences of simple physics.

But how? This is one of the most profound objections to the reductionist way of thinking. It is not at all obvious that a complex world can be constructed from simple premises. If it can't, there is no point in looking for a simple physical basis to our world. We all suffer from severely limited perspective in the cosmic realm: This is the only world we know, and there is much that we don't know about this world.

THE GAME OF LIFE

There is a fantastic computer game called Life. It is played by collegiate computer hackers, by distracted employees of companies with large computers, by filmmakers experimenting with computer animation, and by sundry others on home computers. Life was devised in 1970 by John Horton Conway, a young mathematician at Gonville and Caius College of the University of Cambridge. It was introduced to the world at large via two of Martin Gardner's columns in *Scientific American* (October 1970 and February 1971). The game has had a cult following ever since. Life went through a notorious phase in which it was often played on stolen computer time. (In 1974, *Time* magazine complained that "millions of dollars in valuable computer time may have already been wasted by the game's growing horde of fanatics.") The advent of inexpensive home computers has opened Life to a much wider audience.

Life is described as a game or, sometimes, a video art form. Neither label quite captures the appeal of Life. Certainly Life is nothing like familiar video games. No one ever wins or loses. Life is more like a video kaleidoscope—a way of producing abstract moving pictures on a television screen.

But it's more than that. The Life screen, or plane, is a world unto itself. It has its own objects, phenomena, and physical laws. It is a window onto an alternate universe.

Shimmering forms pulsate, grow, or flicker out of existence. "Gliders" slide across the screen. Life fields tend to fragment into a "constellation" of scattered geometric forms suggestive of a Miró painting.

Much of the intrigue of Life is the suspicion that there are "living" objects in the Life universe. Conway adapted Von Neumann's reasoning to prove that there are immense Life objects that can reproduce themselves. There is reason to believe that some self-reproducing Life objects could react to their environment, evolve into more complex "organisms," and even become intelligent.

Conway further showed that the Life universe—meaning by that a hypothetical infinite Life screen—is not fundamentally less rich than our own. All the variety, complexity, and paradox of our world can be compressed into the two dimensions of the

Life plane. There are Life objects that model every precisely definable aspect of the real world.

Life's rules are marvelously simple. In 1970 Conway was trying to develop a cellular game—a game to be played in the same sort of imaginary universe as Ulam's games. Conway was aware of Ulam's games and of other cellular games inspired by them. To Conway, the most interesting thing about these games was their unpredictability—this in spite of rather simple rules. Conway wanted to create a game that would be as unpredictable as possible, yet with the simplest possible rules.

Conway experimented with many sets of rules. He is said to have devised a game he called Actresses and Bishops. After further thought, he concluded that the rules could be more simplified yet. The simplified game became Life.

Life is a nongame. The rules determine everything so that the game plays itself. Life uses a checkerboard, a sheet of graph paper, or a video screen that is, ideally, infinite. Each cell may be in one of two states. The states are called on and off, 1 and 0, occupied and empty, or live and dead. The cells themselves are sometimes called "bits" (in reference to the two states) or "pixels" (the rectangular picture elements making up a video display). Occasionally, these terms are restricted to cells in the on state: A "three-pixel object" is three on pixels surrounded by off pixels.

Checkers mark on cells on a checkerboard; circles are used on graph paper. On a video display, pixels are light to represent the on state, dark to represent off.

At any instant, the Life universe can be described completely by saying which cells are on. Time flows in discrete moments, digital-clock fashion. The units of time are sometimes called "generations" or "ticks." The situation at one moment determines the situation in the next moment. That situation, in turn, determines the situation in the moment after that. Everything that happens in Life is predestined.

The rules of Life may be handled by a human player or a computer. Each square cell has eight neighboring cells. It adjoins four cells on its edges and touches four more at the corners. During each moment of time, the player or computer counts the number of neighboring cells that are on for each and every cell.

If, for a given cell, the number of on neighbors is exactly two, the cell maintains its status quo into the next generation. If the cell is on, it stays on; if it is off, it stays off.

If the number of on neighbors is exactly three, the cell will be on in the next generation. This is so regardless of the cell's present state.

If the number of on neighbors is zero, one, four, five, six, seven, or eight, the cell will be off in the next generation.

There are no other rules. Conway's name of Life compares the growth of patterns to the growth of populations of living organisms—say, bacteria in a culture. At any rate, the analogy may help you remember the rules. An on cell with fewer than two neighbors dies of isolation. A cell with four or more neighbors dies of overpopulation. Two or three neighbors are just right. If an empty niche has three neighbors, trisexual mating occurs and a new cell is born.

In most games, players make decisions throughout the course of play. In Life, the player's role is almost nonexistent. Life is a sort of spreadsheet program in which each cell's action is dictated by its neighbors. (The game is sometimes played on financial planning software.) The player merely decides what cells are on at the outset—at time 0. Even that role may be abdicated by electing a random assortment of on and off cells. From then on, the inexorable rules of Life determine everything.

BLINKERS, BLOCKS, BEEHIVES, AND GLIDERS

The best way to get the feel of Life is to use a checkerboard or a sheet of graph paper. Conway and colleagues played with black and white checkers. The black checkers marked on cells. Conway identified cells due for a birth and placed a white checker in each. A second black checker marked on cells due to die. When all births and deaths had been identified, the double black checkers were removed from the board and the white checkers were replaced with black checkers. The process was repeated for each new generation. To play on graph paper, make a new diagram for each generation.

Try some simple configurations. If the Life universe starts

out empty—no checkers or on cells at all—every cell has zero on neighbors. By the rules of Life, every cell remains empty in the next generation and the next and the next. Nothing happens.

Take the opposite approach. Start with every cell occupied, an infinity of on cells in every direction. Then every cell has eight "live" neighbors and must die. The Life plane is empty as above a generation later.

Creation must be more subtle. Try a single live cell in an empty universe. Now there are two cases to be considered. The live cell has zero live neighbors, so it must die. Each of the neighboring empty cells has the one live cell for a neighbor. That still isn't enough to make any difference. The configuration dies.

Two adjacent cells also die. Three live cells is the minimum for survival in the Life universe. Try three in a row. The center cell has two live neighbors and survives. The two end cells have just the center cell for a neighbor and die. Above and below the center cell are empty cells that have all three live cells for neighbors. Both cells experience a birth. The result is a column of three cells.

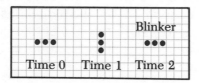

By the same reasoning, the column of three cells reproduces the row the generation after that. The triplet oscillates between the two configurations indefinitely. Conway dubbed this object the "blinker." Blinkers turn up frequently in Life. It isn't necessary to start with a blinker to end up with one.

Not all triplets create a blinker. Three cells in an L shape form a two-by-two square, the "block."

Unlike the blinker, the block never changes. Each of the four cells has three neighbors, so all survive. None of the surrounding empty cells has more than two live neighbors, so there are no births. Conway's term for such stable patterns is "still life." The block is the commonest still life.

A row of four cells produces a different type of still life. The two end cells die but four new cells are born. The result is a two-by-three rectangle. The rectangle is not stable. It in turn changes into a "beehive."

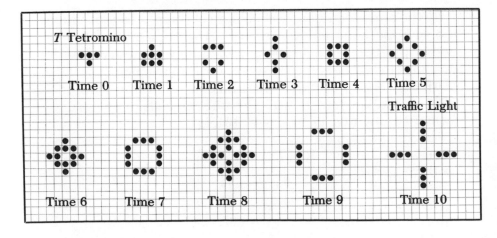

The beehive is a stable hexagon of six cells. Every live cell has two neighbors. The empty cells outside the beehive all have less than three neighbors. The two empty cells inside the beehive each have five neighbors, too many for a birth.

There are many other starting patterns of four cells. The most interesting is called the *T* tetromino. *Tetromino* is an invented word patterned after *domino*. It refers to any pattern of four squares connected by the edges (as the two square halves of a domino are). Both the row of four cells and the block qualify as tetrominos. The *T* tetromino is the T-shaped arrangement of four cells.

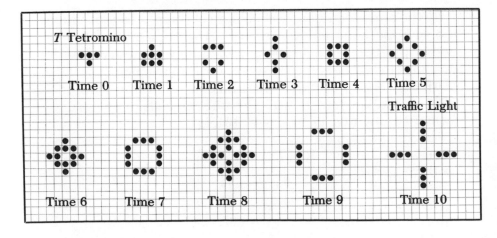

The *T* tetromino is unstable. It expands, becoming a symmetric square at time 4. Generation 5 is a larger square with diagonal sides. The evolution terminates in a symmetric constellation of four blinkers, far enough apart so that they do not interfere with each other. This arrangement of blinkers is itself common enough to have a name: "traffic light." On a fast computer display, a traffic light becomes a twinkling star.

Life is unpredictable. It is surprising that the *T* tetromino should have such a different fate from other small patterns. In one sense, there is no mystery. The *T* tetromino's evolution into a traffic light is demanded by the same count-the-neighbors rule that applies to any other pattern.

Nothing unexpected happens if you focus on any one cell. Look at the center cell of the *T,* where the upright joins the crossbar. It has three neighbors at time 0, so it survives to time 1. Then it has six neighbors, so it vanishes by time 2. Never again does the cell have exactly three neighbors, so it remains empty.

In a bigger sense, there is a mystery. Predictable as Life is on a cell-by-cell basis, the large-scale evolution of patterns defies intuition.

When Life was first introduced, three of the biggest questions Life players wondered about were these: Is there any general way of telling what a pattern will do? Can any pattern grow without limit (so that the number of live cells keeps getting bigger and bigger)? Do all patterns eventually settle down into a stable object or group of objects?

Actually, Conway chose the rules of Life just so that these sorts of questions would be hard to answer. He tried many different numerical thresholds for birth and survival. He had three objectives.

First, Conway wanted to make sure that no simple pattern would *obviously* grow without limit. It should not be easy to prove that any simple pattern grows forever.

Second, he wanted to ensure, nonetheless, that some simple patterns do grow wildly. There should be patterns that look like they might grow forever.

Third, there should be simple patterns that evolve for a long time before stabilizing. A pattern stabilizes by either vanishing completely or producing a constellation of stable objects.

In the real world, matter-energy cannot be created or destroyed. If a colony of bacteria grow to cover a Petri dish, it is only by consuming nutrient. The Petri dish as a whole weighs the same before and after.

No such restriction applies in Life. The T tetromino's four cells grow into a traffic light's twelve. The amount of "matter" (number of on cells) in the Life universe can fluctuate arbitrarily.

If Life's rules said that any cell with a live neighbor qualifies for a birth and no cell ever dies, then any initial pattern would grow like a crystal endlessly. If on the other hand the rules were too anti-growth, then everything would die out. Conway balanced the tendencies for growth and death so precariously that Life is ever surprising.

One of the first surprises was the discovery that some Life patterns move. While following a complex pattern, one of Conway's colleagues noticed that a five-cell unit was "walking." The moving unit was named the "glider."

Glider

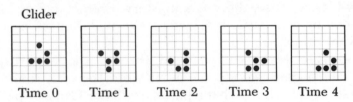

Time 0   Time 1   Time 2   Time 3   Time 4

The glider creeps something like an amoeba or hydra, changing its shape as it goes. It assumes four different phases. Two phases are the shifted mirror images of the other two. Any phase is exactly reproduced four generations later. By then, the glider has moved one cell diagonally.

Mathematicians call a rotated mirror image a glide reflection. That and the suggestion of motion prompted the name. A glider continues on its way forever unless it runs into something. All gliders move diagonally, like bishops in chess. The glider illustrated moves to the southeast, but gliders with other orientations move in the other three diagonal directions. All gliders move at the same speed: one cell diagonally per four generations, or one-quarter cell per generation.

Objects such as the glider motivated Conway to program Life into a computer. The squares of the checkerboard became the pixels of a video screen. On a fast computer display, the motion of a glider is readily apparent.

The most efficient Life programs for home computers typically display two to ten generations a second. The more generations per second, the faster things move. The appearance of gliders and other Life objects varies with speed of display. At a few generations per second, gliders "wag their tails" as they move—the tail being the diagonally connected trailing pixel.

Higher speeds allow different kinetic effects. Some people can see gliders "rotate" on a fast display. Large computers can be programmed to display Life so quickly that gliders zip away, all internal contortions being lost.

The glider is one of the commonest Life objects. When a Life screen starts with a random pattern of on and off pixels, gliders form naturally out of the chaos. Yet Conway did not "put" the gliders into Life. The designers of ordinary video games have to sit down, draw the graphics, figure how to animate them, and write it all up as a complicated program. Life's program is simple and seems to say nothing about gliders (or blinkers, blocks, beehives . . .). Everything you see, no matter how unexpected, is the inevitable consequence of Conway's rules.

Simple rules can have complex consequences. *This* simple rule has such a wealth of implications that it is worth examining in detail. It is the far from self-evident guiding principle of reductionism and of most modern investigations into cosmic complexity. Reductionism will not be truly successful until physicists and cosmologists demonstrate that the large-scale phenomena of the world arise from fundamental physics alone.

This lofty goal is still out of reach. There is uncertainty not only in how physics generates the structures of our world but also in what the truly fundamental laws of physics are. The still-incomplete edifice of physics complicates any study of cosmic complexity.

Five chapters of this book deal with Life. These chapters are not merely about Conway's mock physics; they illuminate the universal issues of complexity and simplicity. The motivation for using Life as an example is that its most fundamental level of physics—Conway's rules—is known from the outset. Our world's is not. The Life player gets to look into another world with phenomena as seemingly arbitrary as our own world's. But the arbitrariness is only on the surface. The Life universe is eminently understandable. The question "Why?" always has an answer, and the answer derives, ultimately, from simple arithmetic: count the neighbors. The Life universe is one of the most vivid examples anyone has found of a complex world that derives from simple premises.

Nothing in this book depends on seeing Life on a computer. All of Conway's arguments and their parallels in the real world are understandable without a computer. All the Life objects discussed will be illustrated. Any object may be followed (for a few generations, anyway) by hand on graph paper. For readers with computers, software advice and programs are included in an appendix beginning on page 233.

·2·

THE LIFE UNIVERSE

The most intriguing five-pixel pattern in Life is the "R pentomino." A pentomino is a five-cell edge-connected pattern. (The glider phases are not pentominos because one of the cells is connected diagonally.) The R pentomino looks a little like the lowercase letter R. If you have a computer, enter the R pentomino and watch what happens.

Phantasmagoric visions appear!

R Pentomino

The R pentomino is wildly unstable. One configuration leads to another and another and another, each different from all of its predecessors. On a high-speed computer display, the R pentomino roils furiously. It expands, scattering debris over the Life plane and ejecting gliders. (Conway's group discovered the glider while tracking the R pentomino.)

The R pentomino's first few generations are unexceptional. It becomes a hook-shaped form at time 1. At time 2, it is a beehive with an extra pixel. Unlike the T tetromino, it never becomes symmetric. At time 9, the R pentomino breaks up into several pieces. But the pieces reassemble into a single eleven-pixel object at time 10.

By time 48, the R pentomino has grouped itself into four objects, two of them familiar. A stable block has formed. The

33

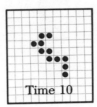

other three objects are still evolving. To the southeast of the block is a traffic-light forming. It is generation 7 of the *T* tetromino, two generations before the first mature traffic-light configuration.

Northeast of the block is another symmetric pattern.

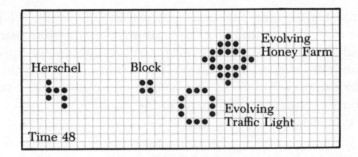

Left to itself, it would become an arrangement of four beehives, called a "honey farm." Like the traffic light, the honey farm has many small predecessors and is often seen.

To the west of the block is a heptomino (seven-cell edge-connected pattern). Conway called this heptomino "Herschel" because it resembles both an *H* and the astronomical symbol for Uranus, the planet Herschel discovered. Herschel is an unstable asymmetric object like the original *R* pentomino.

Both the traffic–light and honey-farm predecessors are growing outward. In isolation, the honey-farm predecessor passes through a succession of symmetric phases, opening out like a four-petaled "flower." The *R* pentomino's honey-farm predecessor does not have enough room, however. At generation 50, it interacts with the traffic light. The traffic-light blinkers are eaten away one by one; the evolving honey farm collapses like a bubble. The symmetry of both patterns is destroyed.

Meanwhile, the Herschel goes through its own evolution

uninterrupted. About twenty generations later, it ejects a glider to the northeast. The glider just clears a diagonal line of three blocks (the original block plus two new ones created from the honey farm–traffic light interaction) and escapes.

The R pentonimo is so prolific that it ultimately fills any low-resolution computer screen with Life objects. Ideally, Life is played on an infinite screen. But a computer, having a finite memory capacity, cannot keep track of an infinity of cells. Life programs usually limit themselves to a finite rectangular field.

Once a growing pattern touches the edge of the field, its evolution may not be the same as it would be on an infinite field. Life programs ignore any growth that would take place beyond the edge. The cells beyond the edge are assumed to be off and stay off for all time. Yet growth that would occur beyond the edge could affect the display area of the field. The evolution of the R pentomino on a small finite field may be nothing like the evolution on an infinite field.

For this reason, Conway did not know the fate of the R pentomino for some time. The inner region remains active after ejecting the first glider. By time 200, four additional gliders have been created and cleared the debris. And no end is in sight . . .

Some Life players wondered if the R pentomino grows forever. Of course the fact that it throws out gliders means that it requires an ever larger checkerboard or video screen to show the complete configuration. But "growth" in Life refers to increase in the number of on pixels. Once ejected, a glider has a constant pixel count of five. To keep on growing, the R pentomino must keep on ejecting gliders or keep filling ever larger regions of the plane with Life debris.

There is no simple way of being sure that the R pentomino does or does not do that. It grows, but there is no pattern to the growth. Parts of the inner region freeze into seemingly permanent configurations, only to be consumed later by still-active regions. The fact that it ejects four gliders from time 100 to 200 does not guarantee that it will continue to do so. No gliders are ejected from time 200 to 300.

The R pentomino was tracked to a final steady state in late 1970. Only at time 1103 does it settle down. It then consists of stable debris that just fits in a 51-by-109-pixel rectangle and, far away, six receding gliders. The sixth and last glider is ejected at about time 700. The diagram does not show the gliders. The position of the original R pentomino is indicated; these five pixels are empty in the steady state.

The R pentomino's final constellation consists of twenty-five objects, including the gliders. Three objects are new still lifes. At the bottom of the diagram is a six-pixel object, the "ship." The ship resembles a beehive, but its long axis is diagonal. The constellation also includes a five-pixel "boat," which is a ship with one of the stern or bow pixels removed. The "loaf" is a common half-circle form of seven pixels.

The object produced in greatest number is the simple block. The census:

> 8 blocks
> 6 gliders
> 4 beehives
> 4 blinkers
> 1 boat
> 1 loaf
> 1 ship

The final population is 116 on pixels. (The maximum population of 319 pixels occurs in generation 821, well before the pattern stabilizes.) Relative to the diagram, the first glider travels northwest, the next three travel northeast, and the last two go southwest.

To say that the R pentomino stabilizes is to say that its future is predictable. The gliders still move, and the blinkers still blink. But nothing in the R pentomino constellation interacts with anything else after time 1103.

The R pentomino's evolution is incredible. Somehow all the blocks, gliders, beehives, blinkers, and other objects are latent in the original R pentomino—but how? The R pentomino is one of many Life objects that defies ready analysis.

STILL LIFES AND THE EATER

The R pentomino creates five types of still lifes. There are more—infinitely many more.

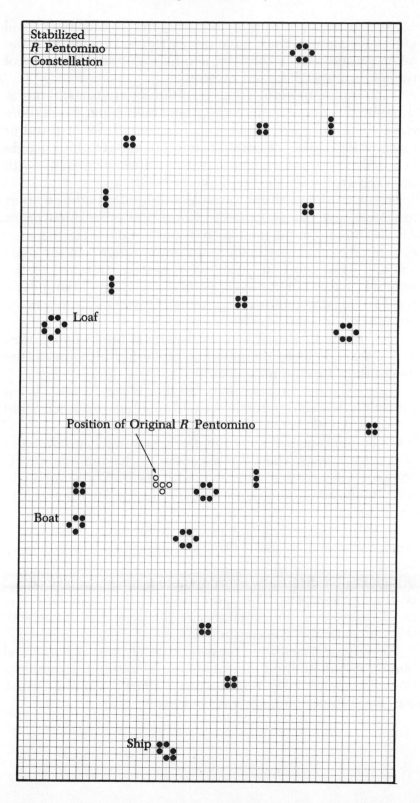

Even so, most patterns are not still lifes. To qualify as a still life, every on pixel must have two or three on neighbors, and no off pixel can have three on neighbors. Otherwise, parts of the pattern would die out or grow.

Stable as the block is, no larger solid square is a still life. A three-by-three square evolves into a traffic light. No still life can have an exposed vertical or horizontal row of three or more pixels—adjacent empty pixels would have three neighbors and experience a birth. Nor can a still life be so dense with on pixels that some die from overpopulation.

The chart shows all the possible still lifes of eight pixels or less. Most have names coined by Conway or other discoverers. All the still lifes of the same pixel count are shown in the same row. Similar still lifes of different sizes are shown in the same column. Thus the four-pixel "tub" can be stretched to the six-pixel "barge" or the eight-pixel "long barge." Barge-type still lifes of any length are possible. Other still lifes can be lengthened too.

The "aircraft carrier" is an unusual still life in two pieces. The L-shaped pieces' neighborhoods overlap and prevent the pieces from turning into blocks.

One of the most remarkable still lifes is the "eater." It has the ability to destroy (totally or partially) nearby objects and yet preserve its own structure.

Other still lifes, including the block, can "eat" certain types of nearby objects. The eater is unusual in having a relatively complex fishhook-shaped region with the ability to repair itself. The eater is the smallest stable object to incorporate this structure.

An eater can eat a glider in four generations. Whatever is being consumed, the basic process is the same. A bridge forms between the eater and its prey. In the next generation, the bridge region dies from overpopulation, taking a bite out of both eater and prey. The eater then repairs itself. The prey usually cannot. If the remainder of the prey dies out as with the glider, the prey is consumed.

There is an attack zone for each type of prey. A horizontal blinker in the position indicated will be eaten in six generations. Eating a blinker requires two "gulps": A bridge forms, dies out, forms again, and dies out again.

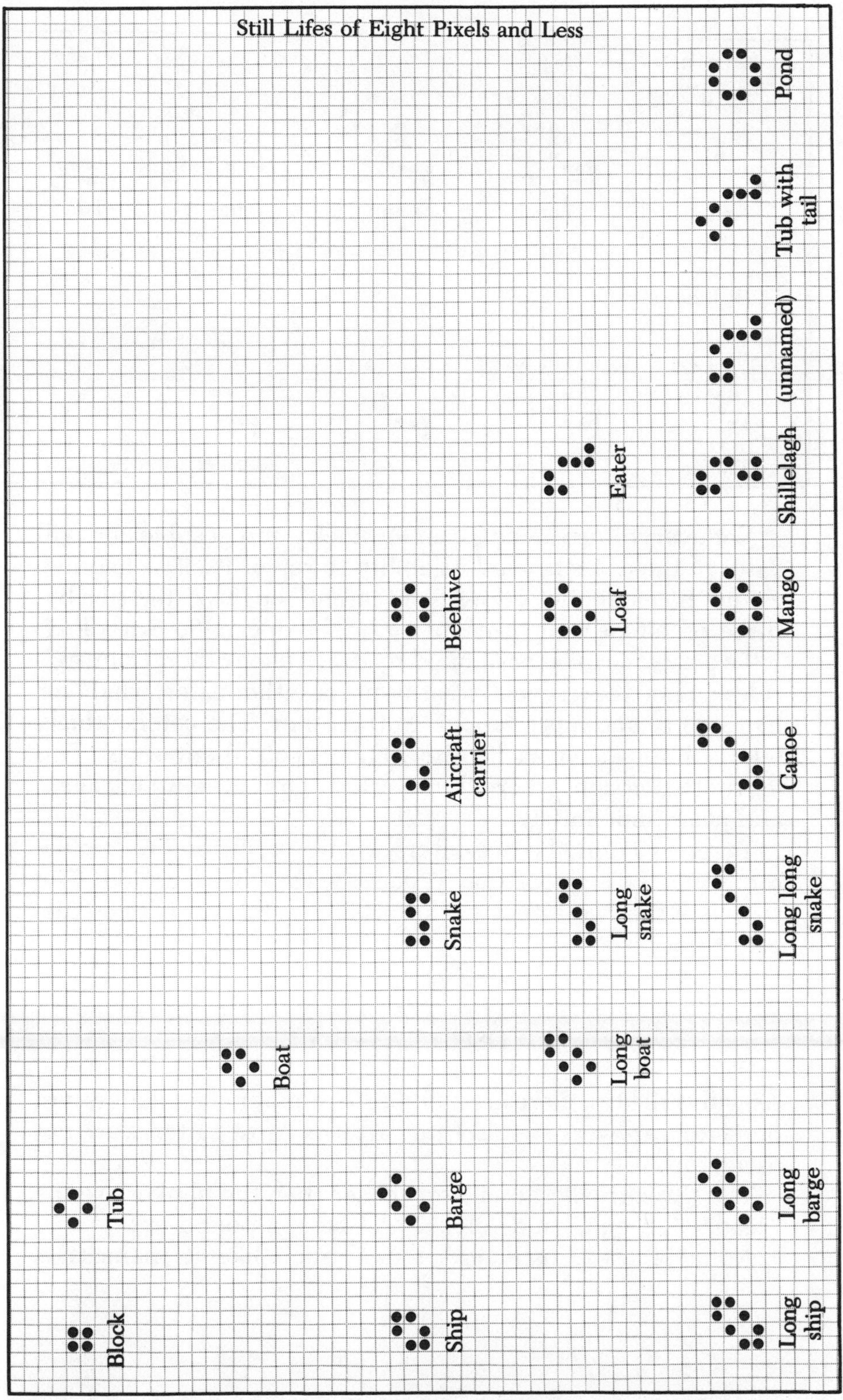

Still Lifes of Eight Pixels and Less

Pond

Tub with tail

(unnamed)

Eater

Shillelagh

Beehive

Loaf

Mango

Aircraft carrier

Canoe

Snake

Long snake

Long long snake

Boat

Long boat

Tub

Barge

Long barge

Block

Ship

Long ship

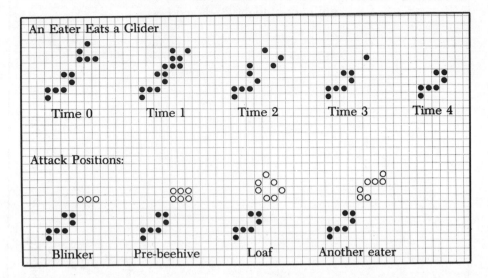

Not everything can be eaten. A block is indigestible and will damage the eater beyond its ability to repair itself. A beehive cannot be eaten, but a "pre-" or "latent" beehive (the two-by-three rectangle that turns into a beehive) can. So can a loaf.

Two eaters can be positioned to attack each other. Conway's rules do not allow special treatment for either eater; whatever happens to one must happen to the other. The eaters attack, take a bite out of each other, and repair themselves. This leaves things right where they started. The eaters attack again and again indefinitely.

Taken as a unit, the two eaters are an "oscillator." An oscillator is any Life object that repeats with a fixed period. The period of the two-eater oscillator is 3. The blinker is an oscillator of period 2.

OSCILLATORS

As with still lifes, there are an infinite number of oscillators, yet oscillators are very special Life patterns. Most patterns are not oscillators.

The three smallest oscillators other than the blinker are the "toad," "beacon," and "clock." All have a period of 2.

None of these oscillators looks much like its namesake.

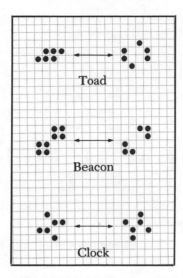

Names of Life objects usually refer to an object's behavior rather than its appearance at any given moment. The toad bloats, the beacon's inner region flashes, and the clock ticks or rotates. The toad and clock have a constant pixel count of 6. The beacon alternates between six and eight on pixels.

It is one thing to understand the cycle of an oscillator on paper; it is often another thing to see it on a fast computer display. Many oscillators produce kinetic effects. The toad seems to puff up and deflate. It may suggest clapping hands.

The clock's two stationary inner pixels provide a frame of reference for interpreting the oscillation of the four outer pixels. The "rotation" of the outer pixels is ambiguous. You can say that the outer pixels rotate 45 degrees clockwise each generation. A given pixel circles the clock in eight generations. But it is equally accurate to say that the pixels are rotating counterclockwise. On a fast display, many people can see the clock as rotating in either sense at will.

Still another way of interpreting the clock is to see the entire structure as rocking back and forth. This perception is closer to the reality because there are, after all, only the two states. Each state is the mirror image of the other along a diagonal axis.

One of the more common oscillators is surprisingly large.

The "pulsar" is a period-3 oscillator whose phases contain 48, 56, and 72 pixels respectively.

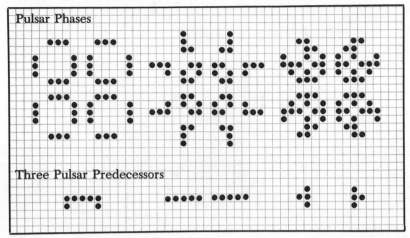

Like the traffic light, the pulsar can arise from patterns much smaller than itself. One predecessor is a seven-pixel object. Another predecessor is a pair of two colinear rows of five pixels separated by an empty space. The pulsar can also form from a pair of T tetrominos that interact as they grow. The two symmetric traffic-light predecessors approach and then split along the line joining them to form four traffic-light predecessors. The four traffic-light predecessors are still packed too closely to form four traffic lights. The result is a pulsar—one phase of which resembles a cramped quartet of traffic lights.

The longer the period of an oscillator, the more complex the visual effects. An oscillator is a sort of abstract video animation, each phase being a frame of the action. Some oscillators of medium to long period grow and contract during their cycle. They are sometimes called "pulsators." An example is the "figure 8."

The figure 8 has eight phases. One phase suggests a diagonal "8" or a scaled-up version of the beacon: two three-by-three squares touching diagonally. The figure 8 is easily tracked through its cycle on graph paper. The phase illustrated gives rise to a perfect rectangular outline. A subsequent phase contains a barge that immediately interacts with the rest of the pattern.

Another type of pulsator is easy to create. A simple row of ten pixels—no more, no less—evolves into the period-15 oscil-

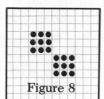

Figure 8

lator called a "pentadecathlon." (The row of ten does not recur but is replaced by the twelve-pixel form shown in subsequent oscillation cycles.) The pentadecathlon is remarkable not only for its long period but for its frenzied history. The minimal phase bloats into a maximum of forty pixels in three generations. It then forms an elliptical ring that instantly disintegrates into two triangles. The two parts evolve separately for a while and then reunite. For eight of its fifteen phases, the pentadecathlon is confined to a narrow zone three pixels wide.

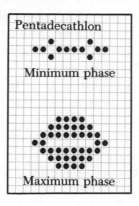

Pentadecathlon

Minimum phase

Maximum phase

The pentadecathlon can also arise from two *T* tetrominos with insufficient room to form paired traffic lights. The spacing between *T* tetrominos must be one pixel less than for a pulsar. Generations 1 and 2 of the *T* tetromino occur in the pentadecathlon (last two phases before the minimal phase recurs).

Oscillators such as the pentadecathlon and pulsar have a special status. They form naturally from small predecessors. Life players soon discovered that it is possible to construct a much greater variety of oscillators than occur naturally. Oscillators may be as large as desired and may display complex kinetic effects.

The "barber pole" is an oscillator that may be stretched diagonally to any length. All the on pixels in the central region have no neighbors and die. A new set of pixels is born to replace them. On a fast display, the barber pole's interior may be interpreted in several ways. The pixels may be seen as moving up or down or in two opposing directions. "Flip-flop" is another oscillator that can be scaled up to any size. Every on pixel dies and is replaced by a birth. Like the clock, it has an ambiguous rotation. Both the barber pole and flip-flop have a period of 2.

"Galaxy" is a period-8 pulsator. The simple geometric phase shown becomes a four-petaled "flower" and then evolves into a spiral "galaxy," throwing off sparks. The fragments reassemble back into the original phase.

The "tumbler's" symmetric halves prevent growth that would otherwise destroy the oscillation. The four quadrants of a pulsar function similarly. The tumbler turns itself upside down every seven generations. The full period is 14. On a high-speed display, the tumbler suggests two snaillike creatures climbing an invisible wall and periodically slipping back down.

The "clock II" is an example of a "billiard-table configuration." This is an oscillator in which the action is restricted to an enclosed region of the Life plane. The clock's bent hand rotates 90 degrees clockwise every generation for a period of 4. The four blocks around the clock II's periphery are necessary to prevent growth on the edges. Many other billiard-table oscillators have been designed.

The period of an oscillator can be arbitrarily long. It is possible for a glider to collide with a pentadecathlon in just such a way that the glider is reflected 180 degrees and the pentadecathlon is unharmed. The glider must encounter the pentadecathlon at the proper phase of oscillation, and the spacing must be right too.

With due attention to details, a glider can be positioned between two pentadecathlons so that it is reflected repeatedly. The pentadecathlons bounce the glider back and forth between them. On a high-speed display, this is reminiscent of a Ping-Pong–type video game, though the paddles are kaleidoscopic.

The shuttling glider is shifted slightly as well as reflected. Its

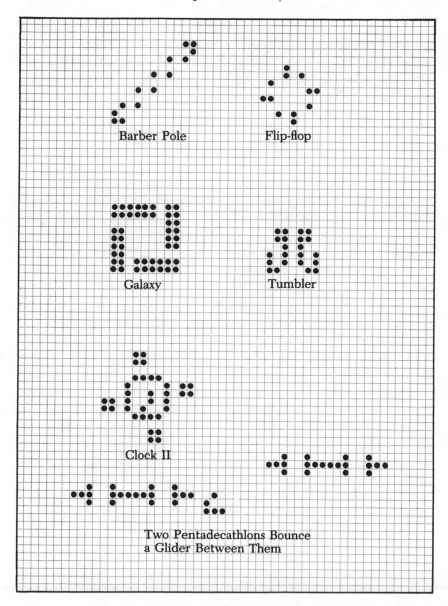

Barber Pole

Flip-flop

Galaxy

Tumbler

Clock II

Two Pentadecathlons Bounce
a Glider Between Them

actual circuit is a narrow parallelogram. The diagram shows the glider in its last normal phase before reflection. The glider, traveling to the southwest, seems almost to have overshot the pentadecathlon. It interacts in the next generation, and then a few generations later. The end of the pentadecathlon that interacts with the glider dies back anyway, so the pentadecathlon is not affected.

The system of two pentadecathlons and shuttling glider is itself an oscillator. Any specific configuration reappears sixty generations later. The diagonal spacing of the pentadecathlons (and thus the period) may be increased freely as long as the phasing works out.

PIXEL ROWS

You may be wondering what a row of nine or eleven pixels creates, insofar as a row of ten gives rise to the pentadecathlon. Conway tracked all the small rows of cells. Surprisingly, the fate of pixel rows depends on the precise number of pixels. One more or one less, and the fate is usually quite different.

One or two pixels die instantly. Three is the blinker. Four becomes a beehive in two generations. The first new case is a row of five. It becomes a solid three-by-three square and then duplicates the later evolution of the T tetromino, terminating with a traffic light.

A row of six dies out in twelve generations. A row of seven is spectacular: It becomes a honey farm via the symmetric display mentioned earlier. The row becomes a circle or octagon in generation 4. Four petals unfurl around generation 8. A central tub (the four-pixel still life) forms the flower's stamens and lasts for several generations. A checkerboard pattern appears in generation 11. By generation 14 it has disintegrated into four beehives—the honey farm.

Why are symmetrical constellations such as the honey farm and beehive so common? Once acquired, symmetry can never be lost under Conway's rules. An evolving pattern that

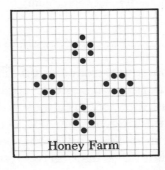

Honey Farm

becomes symmetrical can stabilize only into a configuration that preserves all its symmetry.

A row of eight produces four blocks and four beehives. Nine becomes two traffic lights. Ten is the pentadecathlon predecessor. Eleven becomes two blinkers. Twelve becomes two beehives. Thirteen, like eleven, becomes two blinkers.

A row of fourteen or fifteen cells dies out completely. Sixteen becomes eight blinkers arranged into a "big traffic light." Seventeen creates four blocks. Eighteen or nineteen die out. A row of twenty becomes two blocks.

NATURALISTS AND ENGINEERS

No one knows any simple way of accounting for the behavior of pixel rows, much less for Life objects in general. There are at least two ways of looking at Life. One is the naturalist approach, in which you are interested mainly in seeing what objects occur naturally—what order arises from chaos. The other is the engineer approach, in which you try to construct objects that do complicated, clever, or surprising things.

The definition of a natural Life object is purposely open-ended. Basically, a natural Life object is one that can occur in many different ways. If you can expect to see an object just by playing around with a computer programmed for Life, then it is a natural object.

An object that has many small predecessor patterns is natural. For instance, the abundance of blocks and blinkers in Life is a consequence of the fact that it takes only three pixels to create them. Large objects with predecessors much smaller than themselves (such as the pulsar and honey farm) are more natural than other large objects.

Another way of creating natural objects is to start the Life field in a random configuration of on and off pixels. The initial ratio of on pixels to total pixels (the density) may be any value between 0 and 1. Such random fields ("broths") start out being violently active. Eventually many familiar Life objects— blocks, blinkers, beehives, gliders—appear. The more frequently an object is seen in random fields, the more natural it is.

A third method is to experiment with collisions between gliders or of gliders with other objects. The gliders generated in a random field eventually strike something, so collisions are natural events. Collisions are thoroughly unpredictable. Usually, both objects are destroyed and new objects are created. Objects that occur frequently in collisions are natural objects.

Life is forward-deterministic. A given pattern leads to one, and only one, sequel pattern. Life is not backward-deterministic. A pattern usually has many patterns that may have preceded it. In short, a configuration has only one future but (usually) many possible pasts. This fact is responsible for one of the occasional frustrations of playing Life. Sometimes you will see something interesting happen, stop the program, and be unable to backtrack and repeat it. There is no simple way you can program a computer to go backward from a Life state— there are too many possibilities.

The idea of a natural object can be visualized with a graph. Let circled letters represent Life patterns. Draw an arrow from each pattern to its successor. Because Life is forward-deterministic, there can be only one arrow coming from each pattern. But more than one arrow can point to a pattern, as there are usually many possible predecessors to a pattern. The graph, or a small part of it, will look something like this:

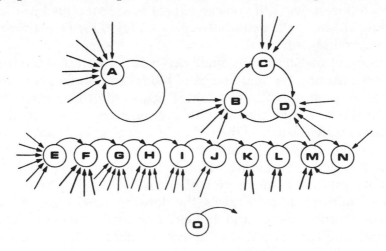

Many arrows point to pattern A, so it is a natural object. The arrow from pattern A also points to itself. A does not change; it is a still life. Pattern A represents a very natural still life such as a block, beehive, or loaf.

Patterns B, C, and D are the phases of an oscillator of period 3. The more arrows that point to the phases, the more natural the oscillator is.

Pattern E represents an unstable object that evolves for some time before stabilizing. It terminates in an oscillating configuration of period 2 (phases M and N). Many other patterns converge on E's evolution (shown by the arrows pointing to the intermediate patterns). So E's final constellation is more common than E itself—much as the traffic light is more common than some of its predecessors.

Pattern O is the loneliest in the diagram. It does not even have itself for a predecessor. It is an unstable pattern with no predecessors. The only way it can possibly turn up on the Life screen is for someone to use it as a starting configuration. The name for such a configuration is a "Garden-of-Eden" pattern.

Alvy Ray Smith III, a mathematician and computer animator, proved that Garden-of-Eden patterns must exist in Life shortly after the game was described in *Scientific American*. Smith did not identify a specific Garden-of-Eden pattern. The first person to do so was Roger Banks. Banks used sophisticated mathematical techniques to prove that a certain 9-by-33 rectangular pattern is a Garden-of-Eden.

This is a pattern with no past. It can never appear in Life except in the initial state. Any pattern that contains this pattern is itself a Garden-of-Eden pattern.

Life is a way of programming a computer to surprise its programmer. The rules are so simple that one feels their consequences should be simple too. This small-scale predictability is important to Life's unpredictability at large scales.

Suppose you set out to write a completely unpredictable graphics program. You might think that the best you could do would be to create video snow—a random pixel pattern such as is seen (more or less) on vacant TV channels.

The trouble is, video snow is predictable by virtue of its very unpredictability. All large regions of the display look just about

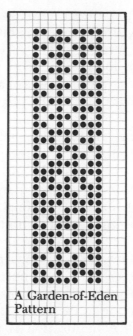

A Garden-of-Eden
Pattern

the same. Even when chance fluctuations create a multipixel dark or light region, it vanishes instantly.

In Life, some of the fluctuations in a random field can linger. Life is predictable enough that many types of structures survive and interact. These interactions give Life a more profound level of unpredictability.

One of the ways in which the naturalist and engineer approaches to Life tie together is in the perplexing question of the fate of an infinite, initially random Life plane. Remember, a finite video screen or sheet of graph paper is a mere window on what is supposed to be a limitless grid. Suppose you start with an infinity of video snow and then apply the rules of Life to it. What happens?

This is the sort of situation that concerns Life naturalists. If attention is limited to a small part of the Life plane (that represented by a computer screen, for instance), it isn't too hard to give an answer. Many on cells die out immediately. The survivors evolve frantically for a long time but ultimately settle into a constellation of still lifes and oscillators.

Is it that simple? Conway wondered about the possibility of patterns that grow without limit. Suppose there is some pattern that grows forever. It needn't even be a natural pattern. *Somewhere* in the tractless infinite plane, every possible pattern—even those that would be considered engineered—must occur.

If they exist, unlimited-growth patterns could change the history of the Life plane. At the very least, they would grow to fill up the empty space available to them. Eventually they would encounter other objects and perhaps the interaction would halt their growth. But a single unlimited-growth pattern might stir up a vast region of the plane.

One can even wonder if some unlimited-growth patterns might be able to preserve their mode of growth while interacting with other Life objects. Such patterns might eventually consume the entire plane.

Evidently, the fate of a small window on an infinite Life plane could depend on the existence or nonexistence of patterns that grow without limit—however contrived or unlikely these patterns may be.

Conway's hunch was that no such unlimited-growth patterns are possible in Life. The tendency to stabilize is too strong, he felt. But Conway conceived of two ways of demonstrating that unlimited growth is possible, if indeed it is.

Both ways involved finding a pattern that does grow forever, yet in a disciplined, predictable manner. One hypothetical pattern Conway called a "glider gun." This would be an oscillator (almost) that creates an escaping glider at every period. The number of pixels would grow by five every period and never stop growing.

Conway called another type of pattern a "puffer train." This would be a moving object, like a glider. As it progressed, a puffer train would leave behind stable objects such as blocks or blinkers.

It is typical of Life's unpredictability that Conway was wrong about unlimited growth. There are patterns that grow without limit. But Conway was exactly right about how unlimited growth occurs. There are glider guns, and there are puffer trains.

·3·

MAXWELL'S DEMON

To appreciate current thought on the origin of cosmic complexity, it is necessary to know something of information theory. For that it is best to start at the beginning. Information theory got its impetus from a perpetual-motion machine that doesn't work, a fabrication presided over by "Maxwell's demon."

James Clerk Maxwell was the Scottish physicist responsible for a set of equations governing all electric and magnetic fields. He was among the first physicists to accept the reality of atoms. In 1871, he published *Theory of Heat.* One brief aside in the book is a fantasy about a remarkable device operated by "a being whose faculties are so sharpened that he can follow every molecule in its course." The device is, in effect, a perpetual-motion machine. It can extract usable energy out of thin air. By 1871 no respectable physicist, least of all Maxwell, thought that such a thing was feasible. Maxwell was posing a puzzle—what is wrong with this machine?

So challenging was this puzzle that the sharp-eyed being became christened Maxwell's demon, and the paradox was not resolved for over half a century. It is still the subject of occasional scientific papers. Let's start with a version of the paradox slightly simplified from Maxwell's original. As background, you need know only that it is not possible to create energy from nothing.

An airtight container is partitioned into two compartments. The wall between the compartments has a tiny trapdoor—so tiny that it is just a little bigger than an air molecule. Maxwell's demon opens and shuts the trapdoor. Whenever a molecule in the lower compartment approaches the trapdoor, the demon opens it and lets the molecule enter the upper compartment. *But* the demon keeps the trapdoor shut when molecules in the upper chamber approach. Molecules from the lower chamber can seep into the upper chamber, but once a molecule is in the upper chamber, it can never get out.

Eventually, Maxwell's demon will sift all the molecules into the upper chamber. There will be a vacuum in the lower compartment and air at twice normal pressure in the upper. Then the demon attaches a U-shaped pipe leading from the upper chamber to the lower. The pipe contains a turbine. A strong wind blows through the pipe, spinning the blades of the turbine until the pressure is equalized. The turbine can mill grain, generate electricity, or perform any other task the demon wants. Where does the energy come from?

CLAUSIUS AND THERMODYNAMICS

Maxwell's demon caused great unease among physicists, since the paradox called into doubt both the first and second laws of thermodynamics.

Both laws had been formulated by German physicist Rudolf Clausius by 1850. Clausius studied the ways that heat energy can be converted into mechanical work. His investigation had as much the flavor of engineering as physics to it. No one really knew what heat was, nor did Clausius's studies delve much into the nature of heat. He simply sought to define some of the limitations that designers of steam engines, for instance, had encountered in converting energy from one form to another.

It was well established that energy takes on many forms, often changing form, both in natural processes and in man-made devices. A fire converts chemical energy into heat; a steam engine converts heat into movement.

Just the same, there were other processes where energy seemed to vanish. The water at the top of a dam has gravitational energy. As it falls, it can power a mill. But if there is no water wheel, the water falls just the same and its energy is lost.

Clausius's first great insight was that energy never really

vanishes, that the total energy of the universe is a constant. This generalization became the first law of thermodynamics. When energy seems to disappear, it only changes form. Look harder and the missing energy will always be there, Clausius argued.

In many cases, missing energy is present as heat. As water falls over a dam, its gravitational energy is converted to motion, and its motion is converted to heat. The water at the bottom of the dam is thus slightly warmer than it would have been otherwise. Had the water powered a mill, it would have been slowed. Less heat would have been created, since some of the water's energy was used for mechanical work.

The first law of thermodynamics also prohibits spontaneous creation of energy. Clausius felt that the failure of engineers to build a true perpetual-motion machine is not a design problem, but an inherent limitation of nature. No amount of ingenuity can get around that limitation. (The term *perpetual-motion machine* is misleading. Nothing prohibits perpetual motion in the absence of friction. The earth can orbit the sun indefinitely. A perpetual-motion machine is something else, a device that creates energy.)

Clausius noted that there are restrictions on the transfer of energy above and beyond the first law. Heat seems to have a unique status among the forms of energy. It is usually easy to convert other forms of energy into heat, but it is difficult— sometimes impossible—to do the reverse.

Clausius drew on the work of French engineer Sadi Carnot. In 1824, Carnot concluded that heat can be converted to mechanical work only where a temperature difference exists. A steam engine at a uniform temperature cannot convert any of its heat energy into motion.

Clausius felt that Carnot's discovery was as fundamental as the first law of thermodynamics. Alone, the first law permitted many processes that simply never happen. The first law would permit water to flow uphill, becoming cooler in the process. It stipulates only that the water's gain in gravitational and kinetic energy be exactly offset by its loss in heat energy.

Any theory of energy transfer that does not forbid such impossibilities is incomplete. In practice, temperature differences tend to even out. They never increase without outside effort. To make this generalization exact, Clausius invented a new mathematical quantity. He first called the quantity *Verwandlungsinhalt,* meaning "transformation content." Later he settled on the shorter name *entropy,* Greek for "turning into."

Clausius's entropy was an abstract quantity residing in physical objects. Whenever heat flowed into or out of a body, its entropy changed as well. The entropy change was defined to be the amount of heat flowing into a body divided by the body's temperature.

Clausius needed the concept of entropy to state his second law of thermodynamics succinctly: The entropy of any closed system can never decrease.

Suppose the closed system is two adjacent reservoirs, one hot and one cold. Heat flows from the hot water to the cold water. The first law of thermodynamics demands that the heat loss of the hot water match the heat gain of the cold. The hot water also loses entropy. The entropy loss is the amount of heat transferred, divided by the hot water's temperature.

The cold water gains entropy. Its entropy gain is the heat gain divided by the cold water's temperature. Although the heat gain equals the other reservoir's heat loss, the cold water's temperature is lower than the hot water's. It follows that the cold water gains more entropy than the hot water loses. The system as a whole has a net entropy gain.

Only processes with a net entropy gain occur in the real world. Notice that a flow of heat from the cold water to the hot would entail an entropy loss—and is forbidden by the second law.

But what is entropy? The second law of thermodynamics refers only to changes in entropy. It is not necessary to define an absolute amount of entropy for any object. Nor is it necessary to interpret entropy as anything more than a heat transfer divided by temperature.

Perhaps the most fruitful early interpretation of entropy was this: Entropy measures the degree to which a system's heat energy is unavailable for doing work. Entropy increases as temperature differences even out. Only systems with a temperature difference can convert some of their heat energy into mechanical work.

ENTROPY AS DISORDER

A more satisfying interpretation would explain why entropy can only increase. Fuller understanding of entropy requires a fuller understanding of heat. Until late in the nineteenth century, the most popular theory held that heat was an invisible fluid that flowed from one object to another. Certain quantitative experiments supported this view, so physicists were reluctant to give it up.

A minority of physicists supported the view (now universally accepted) that heat is the random motion of atoms. The motions of atoms in an object can be broken down into two components. First, there is a motion shared by all the atoms in the object, the observed motion of the object. Second, there is a random atomic motion, different from atom to atom and changing constantly. The aggregate of random atomic motions is what we call heat.

Heat is motion, but it is a collective random motion of trillions of atoms. At any instant, there are atomic motions in every direction. Thus, heat has no unique direction as the motion of a single atom or object does. The higher the temperature, the more vigorous the atomic movement. Temperature measures the average energy of motion of atoms in an object.

The modern interpretation of entropy is due largely to the Austrian physicist Ludwig Boltzmann, one of the most outspoken proponents of the atomic theory of heat. He became depressed and hung himself (in 1906) as scientific opinion seemed to be rejecting his views. Shortly after his death, experimental evidence showed that Boltzmann had been right.

Boltzmann interpreted entropy as disorder at the atomic scale. Prior to this, the most mysterious thing about entropy had been its constant increase. The first law of thermodynamics seemed much easier to accept on an intuitive level than the second. There is only so much energy in the world; it can move around, but not increase or decrease. Entropy, on the other hand, is always being created out of thin air.

Because entropy can only increase, many processes are irreversible. A hot bowl of soup transfers heat to its surroundings and becomes lukewarm. Never, ever, does the reverse happen. A lukewarm bowl of soup cannot absorb heat from its (cooler) surroundings and become hot.

Irreversibility was a novel concept in physics. All the other known laws of physics were reversible, in the sense that if a movie of a process were shown in reverse, the reversal would show an equally possible situation. Newton's gravity, for instance, would allow the planets to orbit the sun clockwise or counterclockwise.

The atomic theory of heat pictures atoms as tiny billiard balls, colliding and rebounding according to the same laws applying to observable collisions. The laws governing idealized frictionless, elastic collisions are reversible. If heat transfer is a mechanical process, why isn't it reversible, too?

Boltzmann realized that there are mechanical processes that are irreversible. One example he used was the mixing of two colors of marbles. A thousand black marbles and a thousand white marbles are placed in a large tray. Initially, the black marbles are on one side of the tray and the white marbles are on the other side. An experimenter shakes the tray. The two colors mix and are soon thoroughly intermingled.

Once the marbles are randomized, no amount of further shaking will ever restore the separation of colors. Shaking can mix the marbles, but it can never sort them.

Any given collision of two marbles *is* reversible, Boltzmann noted. It is only the aggregate of collisions that results in irreversible mixing. The mixing is irreversible in practice rather than in principle. One can always argue that if the tray is shaken long enough, the two colors of marbles ought to separate just by chance. This is so unlikely, however, that it is never observed.

Boltzmann believed the second law of thermodynamics to be a statistical law. Entropy is a measure of the disorder of a system at the atomic level. Individual atomic collisions neither favor nor disfavor order. But the net effect of many random collisions is almost always to decrease order.

Molecules in a hot bowl of soup collide with molecules in the surrounding air. The soup's molecules are traveling faster, on the average, than the air molecules. Any given collision is much like a collision of two billiard balls. When a fast ball hits a ball that is standing still, the still ball almost always gains some of the fast ball's momentum. In consequence, the fast ball slows down. Molecular collisions tend to even out velocity differences too. The soup cools off, warming the air.

Boltzmann felt that Clausius's definition of entropy could be generalized. If entropy is disorder, then its definition should invoke some precise definition of order and disorder.

Disorder is a difficult concept to pin down, and many obvious means of definition fail. It is reasonable to wonder whether disorder can be objectified at all. Disorder seems subjective— in the eye and expectations of the beholder.

Most of the ideas about disorder can be illustrated in a familiar context. Take two offices, one orderly and one disorderly. Both contain the same set of objects: telephone, desk, chair, typewriter, plant, pencil sharpener, etc. Imagine that you must explain the difference to someone who has never seen an office before and has no idea what any of the objects are used for. In what *objective* sense is the disorderly office different?

Boltzmann was faced with much the same problem. Familiar notions of order and disorder are mostly aesthetic. All members of our culture have seen enough interiors to recognize messy ones. Unfortunately, no one has any preconceptions about what a system of atoms is supposed to look like. There is nothing to fall back on. Boltzmann chose to define order in terms of the number of ways that a system may be arranged. The more possible arrangements, the more disorderly the system.

A very orderly office worker has a place for everything and keeps everything in its place. If you inspected the office from time to time, you would always find the same arrangement. In a very disorderly office, on the other hand, any object may be anywhere at all. There are many possible positions for each object and an even greater number of arrangements for the entire office. If you inspected the office periodically, you would not find the same exact arrangement twice.

Boltzmann's definition is statistical. To determine whether an office is orderly, you must observe it a number of times to judge how many possible arrangements it has under its occupant. Nothing can be determined from a single visit. But the definition is totally free of any aesthetic bias. Suppose that the person inspecting the offices comes from a culture where chairs are turned on end, telephones are left off the hook, and all the other furniture is piled in the northwest corner of the room. By coincidence, he finds just such an arrangement on his first visit to the office we think of as disorderly. He might suspect that this office is the orderly one and that the other—which looks to him as if a hurricane hit it—is the disorderly one. After a few visits, though, he would be able to identify the offices as we do. The disorderly office would almost certainly be in different arrangements on subsequent visits. He would conclude that it had no special order and that the arrangement he saw on the first visit was just a fluke. The orderly office would always be in the same arrangement. Whatever he might think of that arrangement, he would conclude that someone was taking pains to preserve it.

In atomic systems, physicists make a distinction between "macrostates" and "microstates." A macrostate is an observed state of a large-scale system: "a liter of oxygen at 1 atmosphere pressure and 0° C.," for instance. A microstate is a precise arrangement of atoms in a system. The molecules in a liter of gas are always moving, so the microstate of such a system is always changing, even when the macrostate does not.

In general, an astronomical number of microstates may correspond to the same macrostate. All the possible arrangements of the disorderly office are microstates. The orderly office is unusual in that it has just one allowed microstate. As atomic systems go, gases have more microstates than liquids, which have more than solids. A given gas molecule bounces off the walls of its container repeatedly and may occupy any position inside the container. Liquid molecules are confined to the volume occupied by the liquid but may be anywhere in this volume. Solids are most orderly. Their atoms are confined to positions in a crystal lattice. Atoms can vibrate about their home positions in a crystal but cannot stray far (when they do, the crystal melts).

The entropy of a macrostate is defined by its number of microstates—the number of arrangements. It might seem that this number should always be infinite, since even a single atom can occupy an infinite continuum of positions. Boltzmann solved this problem by breaking down the possible positions into discrete ranges for the purposes of calculation. He found it easier to deal with the logarithm of the number of microstates than with the number itself. Boltzmann's equation for entropy defines it as a constant times the logarithm of the number of microstates. The use of logarithms is a computational refinement that need not worry you. Entropy is disorder, and disorder has to do with the number of possible arrangements.

Boltzmann's view of entropy is reductionism at its purest. It shows that the second law of thermodynamics is not a fundamental law but can be reduced to the statistics of atoms obeying the simpler laws of motion. At the heart of Boltzmann's thesis is this realization: Practically all the microstates for a complex system are "random" microstates. They are microstates with no evident pattern. Atomic positions and velocities are so evenly distributed that large-scale properties such as temperature and pressure are uniform throughout the system.

A system with no temperature or pressure differences is said to be in thermodynamic equilibrium. The reason that systems tend to thermodynamic equilibrium is simply that the vast majority of microstates are states of thermodynamic equilibrium.

Entropy tends to increase for the same reason that offices tend to get cluttered. Consider all the geometrically possible arrangements of the objects in an office. It is easy to see that there are an astronomical number, even if it is required that positions of objects must differ by at least an inch to count as a distinct arrangement. 99.999 + % of these arrangements look disorderly. For instance, it may not make sense to have the chair anywhere but under the desk, yet obviously most of the possible positions for the chair are not under the desk. Similarly, most of the possible positions for the telephone are not on the desk, and most of the possible positions for the files are not in the file cabinet (and certainly not in alphabetic order!). The arrangements perceived as orderly are far, far outnumbered by those perceived as disorderly. During the day, various objects are shifted inadvertently. Arbitrary shifts are far more likely to produce a cluttered microstate than another orderly one.

In a sense, it's semantics. Out of the universe of possible arrangements, we choose a subset and call them orderly. Everything else is called disorderly. The tendency for offices to get cluttered is in direct proportion to the narrowness of our definition of order.

Even so, the tendency to clutter is more general than any one definition of order. A culture in which desks are supposed to be overturned and furniture is piled in the northwest corner of the room would need cleaning people as much as we do. They too would find that incidental shuffling of a room's arrangement would almost surely put it outside their definition of order.

Physicists can avoid most of the subjectivity of disorder. In atomic systems, only a few large-scale properties such as temperature and pressure are readily measurable, so physicists deal almost exclusively with macrostates that can be defined by these observables. One such macrostate is the one Maxwell's demon tries to produce: a container with air at 2 atmospheres pressure in one half and 0 pressure in the other half. Were it not for the wall between the halves, the compressed air would rush into the vacuum. The reason is not that nature abhors a vacuum. Individual molecules have no way of knowing that there is a vacuum in the other end of the container—they merely bounce about aimlessly in accordance with the laws of motion. But of the astronomical number of ways the molecules might be arranged, virtually all are states of maximum entropy and thermodynamic equilibrium. Shortly after removal of the partition, the odds favor a microstate with 1 atmosphere pressure throughout. The odds are so overwhelming, in fact, that a compressed gas is never observed to *fail* to expand into a region of lower pressure.

HEAT DEATH

Thermodynamics was the first inexact branch of physics. The details of the atomic world are never known (as we will see). Yet is it possible to make predictions about the behavior of systems of atoms. These predictions are never certain, though in many cases (such as gas occupying a vacuum) they are certain enough for any practical purpose. Thermodynamic analysis has made it possible for cosmologists to realize much of Laplace's dream of surmising the past and future of the universe without inquiring into the details of even a single atom's position or velocity.

Thermodynamics seems incapable of explaining the complexity of the world, however. Boltzmann puzzled over the contradiction of a rich, ordered universe and the second law of thermodynamics. The universe is a closed system by definition. It is also the largest possible system. Never can the second law of thermodynamics be more statistically certain than when applied to the entire universe. The entropy of the universe must increase irreversibly every second.

Boltzmann's generation of physicists was haunted by the specter of *Warmestod,* heat death of the universe. It was predicted that all forms of energy would ultimately degrade into heat and all temperature differences would even out. Then entropy would be at a maximum.

The heat-death universe would contain just as much energy as it does now, but all the energy would be in an unusable form. No large-scale processes would be possible. Nothing interesting could happen. The heat death would entail the extinction of all the stars, for stars are concentrations of energy. It would mean the end of all life.

MAXWELL'S DEMON REBUTTED

The paradox of Maxwell's demon plays with the idea that intelligent beings can evade the second law of thermodynamics. A human being can sort a mixture of black and white marbles back into two single-color piles. It seems reasonable that a microscopic demon could do the same with molecules. Maxwell proposed a demon sorting molecules by temperature, not pressure as above. The demon opens and closes the trapdoor so that the relatively fast molecules are sorted into one chamber and the relatively slow molecules in the other. This should result in a temperature difference between the two chambers. In principle, the temperature difference could be harnessed by a steam engine. The paradox is not fundamentally different whether the demon sorts by temperature or pressure. In both cases, the demon's actions decrease the entropy of the system. Let's look at some of the critiques of the demon.

The situation is too unrealistic to be relevant.
There are no atom-sized demons, so why go any further? Many physicists felt that Maxwell's fantasy did not really pose

any threat to thermodynamics. You can imagine a demon that makes objects fall up but that doesn't change the laws of gravity.

Could the demon be replaced with an automatic device?
Apparently, yes. So not all physicists were quick to dismiss the demon as a fiction. The pressure-sorting demon is easier to simulate mechanically. Let the trapdoor be hinged so that it opens only into the high-pressure chamber. Molecules can push but never pull a door, so the molecules in the high-pressure chamber will not be able to open the trapdoor. A spring holds the trapdoor shut normally. Tension in the spring is adjusted so that molecules hitting the door from the low-pressure side can open the door and enter, the spring slamming the door shut behind them.

This is a simple model of an automated demon. Dennis Gabor and Léon Brillouin have devised more sophisticated ones. Anything the demon can do, a machine can be built to do (although this was not nearly so evident in Maxwell's time). The problem must be in the sorting operation itself and not in the unreality of the demon.

The demon does not violate the first law of thermodynamics.
At first sight, the demon's perpetual-motion machine seems to challenge both the first and second laws. A little further thought shows that the first law is inviolate—and on this, all physicists agree.

The demon's actions do not change the basic laws of motion for any one molecule or for the aggregate. In every collision, the energy of the colliding molecules neither increases nor decreases. Take the version where the demon sorts by pressure. The demon is packing a volume of gas into half its initial volume. But he does this, ideally, without exerting any force on the molecules. Each of the molecules is still traveling at the same average velocity as before the compression. The temperature of the compressed air has not changed.

In order to benefit from his efforts, the demon must allow the compressed air to expand. The expansion of the air spins the turbine. It also lowers the temperature of the air, as the molecules transfer some of their velocity to the turbine. The demon does not really end up with what he started with. After he has extracted energy from his machine, the air is at a lower temperature.

No problem. The demon can wait for the air to warm up and then start over. Maxwell's demon operates a "perpetual-motion machine of the second kind." It converts heat energy into work, in violation of the second—but not the first—law of thermodynamics.

Does the demon's sorting require energy?
If it did, it would explain everything. Entropy can always be decreased locally by an outside expenditure of energy (and concurrent entropy increase). This is what a refrigerator does. It increases the temperature difference between its interior and exterior, at the expense of the energy it takes to run it. If the thermodynamics of generating the electric power for the refrigerator are taken into account, the net entropy always increases.

Opening and closing the trapdoor takes energy. This energy must be subtracted from any generated from the temperature or pressure difference created. Now here the simple spring valve described above looks suspicious. In order to get into the high-pressure chamber, each molecule must push against the force of the spring. That slows the molecules slightly, so the molecules making it into the high-pressure chamber must have a slightly lower effective temperature. We might decrease the tension in the spring, but then the door would take longer to slam shut and some of the molecules in the high-pressure chamber might escape in the meantime.

We conclude that such a simple spring valve has problems. Let's go back to the original demon. He merely watches the molecules and, at the right times, opens the door. It is not necessary that the door be on any kind of spring. The trapdoor can be as lightweight and delicately balanced as desired. The demon can open the trapdoor for incoming molecules in advance so no incoming molecule even touches the door. Then the energy expended in operating the door may be arbitrarily small and need not cancel out the energy generated. The crucial point is that the demon must know in advance when to open the door, so that the molecules do not have to push the door open themselves and expend energy. The demon can still be automated, say with an electric eye to detect molecules.

The trapdoor is subject to Brownian motion.
In 1912, M. von Smoluchowski suggested that the problem is Brownian motion, the random jostling of small objects subject to molecular collisions from all sides. This is not exactly wrong, but neither does it much clarify the issue. If the trapdoor is indeed big enough that molecules are likely to be hitting it from both sides at once, then the demon cannot safely open the door anyway. The paradox requires that the door be small enough that the demon can deal with the collisions on a molecule-by-molecule basis. The collisions are Brownian motion, but if the demon sees the molecules coming he can use this Brownian motion.

The demon is foiled by quantum uncertainty.
This was suggested by J. C. Slater in 1939 and is often mentioned in connection with the paradox. Maxwell did not know it, but atoms (and subatomic particles) do not behave like tiny billiard balls. In certain situations they can act like waves rather than particles. Heisenberg showed that this mandates an inherent uncertainty in any simultaneous measurement of a particle's position and momentum. The more precisely the particle's position is known, the less precisely its velocity can be known, and vice versa. The smaller the particle, the more overwhelming this uncertainty becomes.

In order to operate his trapdoor, the demon must measure the position and momentum of molecules near the door. He need not know the molecules' trajectories precisely. If need be, he can open his trapdoor when his measurements merely indicate it is *likely* a molecule will enter from the right direction. In that case, mistakes are inevitable. Sometimes the demon will open the door when he shouldn't and lose ground; other times he will fail to open the door and miss an opportunity. The demon's sorting operation will be slowed—but not stopped. As long as the demon can learn *something* of the molecules' positions and velocities, the odds will favor him in the long run.

Quantum uncertainty makes the demon "near-sighted."

The smaller the molecules he tries to sort, the worse his myopia is. But never is the demon totally blind. Moreover, there are gas molecules massive enough that quantum uncertainty is unimportant at low pressures (uranium hexafluoride, for instance). The demon's real problem lies elsewhere.

A COUNTERFEITING DEMON

Let's consider an economic Maxwell's demon—a situation that seems to bear no resemblance to the original paradox. The demon sets out to attain total wealth by acquiring everything in his home kingdom: all the land, all the houses, all the cars, all the crops, everything. He doesn't actually want to steal from anyone, but neither does he want to do any work. He decides to counterfeit. He fashions duplicates of the kingdom's currency plates so precise that not even treasury officials will be able to tell the difference. That way, no one will be cheated, the demon rationalizes: The people accepting his counterfeit bills will never know it and will be able to spend them as freely as real money. The demon estimates the fair market value of all the goods in his country as one billion of the local dollars. He prints up a billion dollars and starts out on a spending spree.

The demon's counterfeiting is a financial perpetual-motion machine. It, too, is doomed to failure. Let's see why.

The demon will get full value for his first bogus bill. He may be able to spend millions before any problem is evident. It is the megalomaniacal scale of his counterfeiting that will defeat him. The demon has printed a dollar for every dollar's worth of goods in the country. The amount of money he has counterfeited can be no less than the amount of legitimate money in circulation. It follows that the effects of the demon's actions on the economy cannot be ignored.

The demon's counterfeit bills will flood the money supply, throwing the kingdom into a galloping inflation. The demon will be forced to pay ever higher prices. The more money he spends, the more prices will go up. The demon's billion dollars will run out long before he acquires everything.

Of course the demon can always print up more money. Eventually, practically all the money in circulation will be the demon's counterfeit bills. The demon will have to fill wheel-barrows full of $100,000 bills (the highest denomination) just to buy a loaf of bread. Still, the printing power of his press is unlimited.

To print ever greater quantities of money, the demon will need ever greater quantities of paper and ink. One day the demon will push a wheelbarrow of $100,000 bills to the paper and ink store to replenish his supplies. On the way home, he will find that the wheelbarrow, now containing paper and ink, is lighter. If the paper and ink weigh less than the money used to buy it, the demon cannot possibly print enough money to cover its cost!

This is a problem with any sufficiently massive counterfeiting scheme. Eventually, the counterfeiting must inflate the currency to the point where money isn't worth the paper it's printed on. Then further counterfeiting does no good. You actually *lose* money.

INFORMATION AND ENTROPY

The troubles of the molecule-sorting demon are cut from the same cloth. It turns out that the demon's sorting has effects on the system that cannot be ignored and must ultimately confound the demon.

Leo Szilard, a colleague of Von Neumann at the University of Berlin, offered the first modern analysis of Maxwell's demon. Szilard critiqued the demon in a 1929 paper, "Über die Entropieverminderung in einem thermodynamischen System bei Eingriffen intelligenter Wesen" (*Zeitschr. f. Physik* 53, 840–856). His way of looking at the paradox foreshadowed the information theory developed twenty years later by Claude Shannon. An experimenter's information about a system is a vital part of the system, Szilard found. There is no such thing as an omniscient, aloof observer. The act of making an observation changes the system observed. As Von Neumann commented, "An observation is an irreversible process."

Heisenberg's quantum uncertainty is often couched in similar phrases. It is important to understand that quantum uncertainty is distinct from the type of information-theoretic limitations that foil Maxwell's demon. The latter limitations would apply even in a universe where atoms were the tiny billiard balls imagined in classical physics.

Szilard and those who continued his work exposed the folly of aspiring to total knowledge of anything. Not even the position of a single gas molecule can ever be known with total precision. Observation of a system drives up the entropy cost of more precise observation. There comes a point at which any observation is choked off—it ceases to yield enough information to pay for itself.

Szilard pictured a simplified Maxwell's demon experiment in which a cylindrical chamber contains a single gas molecule. The molecule rebounds off the walls of the chamber. Its complicated path traverses all parts of the cylinder's interior. Clearly, the molecule is *somewhere* at any instant. Suppose it is in the top half of the cylinder and that the demon has determined this fact through some observation. Then the demon need only insert a frictionless circular piston in the middle of the cylinder. This confines the molecule to the space above the piston. The molecule will bounce off the piston repeatedly, each time transferring a little downward momentum to it. In effect, the molecule is a "gas" expanding into the vacuum below and pushing a piston as it does so. The (very, very slow!) downward motion of the piston can be used to do work. Eventually, the molecule will push the piston all the way down to the floor of the cylinder. Then no more energy can be extracted.

The energy has a price, Szilard argued. Originally, the demon knew the molecule to be in the top half of the cylinder. Afterward, he knows nothing about the molecule's position. It could be anywhere in the cylinder's volume. The demon has gained energy but sacrificed information.

To repeat the process, the demon needs more information. He needs to observe the molecule's position again. This is such a vital link in Szilard's argument that it deserves emphasis. You might wonder why the demon cannot simply replace the piston in the middle of the cylinder.

If the demon slides the piston back to the midway point, he will be forcing the piston against the effective pressure of the single-molecule gas. This requires energy—the *exact* amount of energy he just extracted. No good.

Instead, the demon will have to remove the piston from the bottom of the cylinder. Perhaps the bottom screws off and the demon can take the piston out quickly, before the molecule escapes. Perhaps, too, there is a slot in the middle of the piston so the demon can slide the piston back in. In that case, the demon will not have to move the piston against any resisting force and expend energy. (The demon's exertions while outside the cylinder are assumed negligible.)

The demon has not observed the molecule, so he has no way of knowing whether it is in the top or bottom half when he replaces the piston. He consequently does not know if the piston will move up or down from the midpoint. Say the piston is hooked up to a flywheel. The original downward stroke of the piston has set the flywheel moving (very, very slowly) clockwise. There is a fifty percent chance that the piston will move up this time and slow the flywheel back to a halt. In the long run, there will be as many downward strokes as upward strokes, so the demon cannot expect the flywheel to pick up any speed in either direction.

It should be possible to solve the problem with two flywheels. An electric switch is tripped when the piston starts moving down, and it engages the piston to a clockwise flywheel. If instead the piston starts moving upward, a different switch engages it to a counterclockwise flywheel. In the long run, both flywheels will move faster and faster and can be used to perform useful work. *But* the system of piston and switches is nothing less than a device for measuring the approximate position (top or bottom half) of the molecule. We conclude that the demon requires information about the molecule's position to make his perpetual-motion machine work.

Maxwell assumed his demon could know molecular positions and velocities to any necessary degree of precision, but Szilard answered that this presumed omniscience is artificial. In real life, knowledge is always the result of observation. Szilard and such physicists as Gabor and Brillouin examined how the demon might attain knowledge of the molecule's position.

Case 1: location by constraint.
Szilard's paper considered the simplest kind of "observation," one so crude that it may seem to have little relevance. Picture Szilard's system of a single molecule in a cylinder. The simplest

question the demon can ask about the molecule's location is whether it is in the top half or bottom half. One way to settle the matter is to place a piston on the floor of the cylinder and raise it to the midpoint. Then the molecule *has* to be in the top half. Or the piston can be placed on the ceiling of the cylinder and lowered, forcing the molecule to be in the bottom half.

This is the same procedure that was mentioned earlier and rejected on the grounds that it requires energy. It seems nothing like ordinary observation. The demon doesn't find out anything about the molecule's original position. Rather, he constrains the molecule to one section of the cylinder in order that he may know the molecule is in that section.

Surprisingly, this type of observation is not fundamentally much different from other types. Szilard chose it for analysis because it is simple. Not only does location of the molecule cost energy, he showed, but also increased information about the molecule's position is always canceled out by an increase in entropy. Entropy, in turn, is ignorance at the molecular level.

Let the demon insert the piston at the bottom of the cylinder and raise it to the halfway point. In effect, he is compressing a very, very thin gas to half its volume. The rebounds of the molecule against the piston resist its upward motion. At the same time, the piston imparts some of its momentum to the molecule. The molecule picks up speed, which is to say its effective temperature increases. This is nothing more than the familiar increase in temperature when a gas is compressed.

For simplicity, Szilard supposed that the cylinder's walls and surroundings conduct heat so efficiently that the molecule is kept at the same effective temperature throughout the compression. Furthermore, the surroundings are so much bigger than the cylinder/molecule system that they can absorb all heat generated without themselves experiencing a measurable increase in temperature.

In that case, all the energy supplied to the piston is transferred to the molecule and converted to heat. The heat energy flows into the cylinder's surroundings.

Whenever heat flows there is a change in entropy. The energy needed to move the piston against the single-molecule gas pressure is readily calculated. This energy is equivalent to a certain amount of heat. Take this heat, and divide it by the cylinder's constant temperature. The result is the cylinder's entropy decrease. The same amount of heat flows into the cylinder's surroundings, which are at the same temperature. So the surroundings' entropy gain equals the cylinder's entropy loss.

The demon has succeeded in decreasing entropy in the cylinder. He knows more about the molecule's position than he did. Before inserting the piston, the molecule was equally likely to be anywhere in the cylinder. It could have any of a range of velocities determined by the temperature. After the demon's action, the range of velocities is the same because the temperature has not changed. But the number of possible positions is cut by half. The total number of possibilities (positions and velocities) for the system is halved.

The cylinder's surroundings include the entire universe outside the cylinder. We know that the surroundings' entropy gain equals the cylinder's entropy loss (Clausius's definition, heat divided by temperature) and that the cylinder's entropy decrease represents a decrease by a factor of two in the number of possibilities (Boltzmann's definition, in terms of the number of possible arrangements). It follows that the number of possible arrangements of the surroundings—for the whole universe, in fact—must increase by a factor of two.

This is the heart of Szilard's thesis. Any increase in knowledge about the lone molecule is (at least) exactly compensated by increased ignorance about the rest of the world. If the demon raises his piston 99 percent of the way from the bottom of the cylinder to the top, he is a hundred times surer about where the molecule is. But the operation produces heat, which is absorbed by the surroundings and which increases the range of possible positions and velocities of molecules in the outside world by a factor of a hundred.

This startling result will be examined more fully in a moment. Let's first look at some of the other ways the demon could observe molecules. Szilard's situation seems too narrow —surely the problem is with the demon's ham-handed way of "observing." Maxwell had his demon watch the molecules, but the notion of seeing a single molecule turns out to be ambiguous, so we will consider three cases of vision at the molecular scale.

Case 2: seeing by ambient light.
All matter radiates, the frequency of radiation depending on the temperature. For objects at room temperature, the frequency is in the infrared range. So if the demon's eyes are sensitive to infrared light, he will find the interior of his sealed chamber bathed in a warm glow. Infrared photons (the fundamental "particles" of light) will bounce off the molecules and into the demon's eyes. He will see the molecules.

This idea is easily dispensed with. Vision does indeed involve photons bouncing off objects and into an observer's eyes, but just as important as the photons themselves is the directional information they carry. The sun is in one part of the sky. If the eye detects photons coming from other directions, then something must be reflecting photons from those directions.

No such inferences are possible for the demon. Suppose first that the chamber contains no molecules, just the infrared photons. The photons are *black-body* radiation, which means that they are in thermodynamic equilibrium. Photons are equally likely to be found in any part of the chamber, traveling in any direction. The demon sees photons coming from all directions equally.

How would the presence of a single molecule in the chamber change things? It wouldn't. Some of the photons the demon sees may have bounced off the molecule. The interaction with the molecule would have changed the photons' course. But in the utter absence of prior knowledge about individual photons, the demon can draw no conclusions. With the molecule or without, photons are equally likely to come from any direction. The demon cannot even identify photons that have interacted with the molecule by a change in frequency, for the black-body photons are at the same effective temperature as the molecule and the chamber walls.

The demon's chamber can hold two molecules or a quintillion. It will still make no difference. Black-body photons' "message" is already as garbled as it can be. Further garbling by molecules will be undetectable. All the demon will ever see is a random snow of photons.

Case 3: using a flashlight.
To see anything, the demon must provide his own light. It must be a coherent beam—light whose initial direction is known so that a change in direction (by reflection off a molecule) is detectable. The filament of the demon's flashlight must be at a higher temperature than the chamber so that reflected photons will not be confused with the black-body radiation from the chamber itself.

This is a plausible way of gaining information about molecular positions and velocities. Again, there is a catch. Brillouin calculated the entropy change associated with observation. Observation of a molecule's position requires (at least) one photon from the flashlight striking the molecule. The photon transfers heat from the hot filament to the molecule and thus to the chamber and its surroundings. As in any spontaneous heat flow, there is an increase in entropy.

Brillouin's calculations showed that the entropy gain must at least equal, and usually exceed, the information the demon obtains. Observation by flashlight has the same fundamental problem as observation by Szilard's piston. In fact, Brillouin argued, if the demon wants to create a temperature difference between two compartments of gas, he can do no better than to forget sorting and use his flashlight filament to heat up one compartment.

Case 4: seeing with "sub-quantum" particles.
The problem with a flashlight seems trivial. It so happens that the particles we use for seeing (photons) have "sizes" and energies comparable with the molecules'. Maybe the problem would go away if smaller particles were used for imaging.

All the known subatomic particles are bound by Heisenberg's uncertainty principle. The resolution of any microscope using photons, electrons, or other quanta is limited.

One can imagine a new particle or type of particle, much smaller than the known subatomic particles. Here "smaller" means a smaller mass and effective diameter. Call these hypothetical particles sub-quantum particles and assume that they are not subject to quantum uncertainty. Then sub-quantum particles could form the basis for a high-resolution microscope that would allow the demon to observe and sort molecules.

The microscope would shoot a coherent stream of sub-quantum particles at molecules to reveal their positions and velocities. The sub-quantum particles would transfer energy to the molecules, resulting in a slight transfer of heat to the system. But the heat transfer depends on the momentum of the sub-quantum particles, and we'll assume that this momentum is negligibly small.

What's wrong here? Nothing, really. *If* there were sub-quantum particles as described, presumably they could be used to sort molecules. The sorting of molecules observed with sub-quantum particles would not be much different from the sorting of black and white marbles observed with photons.

In a larger sense, sub-quantum particles would only transfer the Maxwell's demon paradox to the sub-quantum realm. One would end up wondering whether a sub-demon could sort sub-quantum particles. The sub-demon would run into the same problems the demon does.

Maxwell's demon must be assumed to operate at the lowest level of matter and energy. His difficulties are broader than any particular brand of physics. They have to do with information. If the demon can learn about particles and waves only by observing them with other particles and waves, then never can he come out ahead.

THE LIMITS OF EMPIRICAL KNOWLEDGE

The moral is that there may be a great difference between omniscience and knowledge that must be obtained through observation. The distinction is tantalizing. For any given two-chamber container of gas, there is clearly some sequence of openings and closings of a trapdoor that will sort the molecules. The sequence is real and could be written out: "40 nanoseconds open; 12 nanoseconds closed; 350 nanoseconds open; . . ." If *only* the demon could know this sequence, his perpetual-motion machine would work. Yet any attempt to learn the sequence defeats the demon in the long run.

Szilard's cylinder-piston single-molecule version of the paradox is a good way of dramatizing the limitations of empirical knowledge. The demon aspires to know the exact vertical position of the molecule. He inserts the piston in the top of the cylinder. Initially, the molecule may be anywhere in the cylinder's volume. The demon pushes the piston halfway down to the bottom. Now his uncertainty about the molecule's position has been decreased by a factor of two.

Since the piston offers virtually no resistance, the demon pushes it 99.99999999999999999999 percent of the way to the bottom. This confines the molecule to a wafer-thin volume. The demon's uncertainty about the molecule's vertical position is decreased by a factor of 10,000,000,000,000,-000,000,000.

Now the piston is starting to push back. The molecule (which is of an ideal gas) has about as much room as the volume per molecule in a gas at 1-atmosphere pressure. The molecule bounces between the piston and the floor so frequently that it acts like a gas at ordinary pressure. The compression generates heat. Now the bottom of the cylinder is noticeably warm.

Has the demon increased his total knowledge of the world? Szilard's analysis says no, in this sense: The demon is now at least 10,000,000,000,000,000,000,000,000 times less sure of the exact state of the universe outside his cylinder.

The culprit is the heat produced. The demon had to expend mechanical energy to compress the gas. This requires that the atoms in the demon's arm all move in the same direction. In return for this energy, the demon gets back heat. Heat is chaotic motion of atoms. The hotter something is, the less certain one is of the velocities and positions of its atoms.

To say that the demon is 10,000,000,000,000,000,000,000,000 times less sure about the outside world means that the number of possible arrangements of atoms in the surroundings warmed by the compression has increased by this factor. This is more difficult to picture than, say, the number of arrangements of an office, but it is the same idea.

At the beginning of the day, a wastebasket is located just right of the desk, somewhere in a four-square-foot area. At the end of the day, it has been moved around so much that it is equally likely to be found anywhere in the office's one hundred square feet of floorspace. Then the number of possible arrangements of the office has increased twenty-fivefold. So has the number of possible arrangements for the entire office building, or indeed (speaking in somewhat oversimplified terms) the entire universe. We do not have to wait for the heat produced by the demon's compression to diffuse to the distant galaxies. Once the cylinder and surroundings have cooled down to their initial temperature, it can be demonstrated that the number of atomic arrangements for the universe exclusive of the trapped molecule is 10,000,000,000,000,000,000,000,000 times greater.

The demon will find himself unable ever to attain total knowledge of the molecule's position. As the piston is pushed to the floor, the molecule's effective pressure increases without limit. Soon the demon's arm will give out and he will have to resort to a hydraulic press or other device. Compression becomes ever more expensive in terms of energy and entropy. Eventually, the demon will need a refrigeration system just to prevent the cylinder from melting. Not all the energy on earth will suffice to squash an ideal point-sized molecule to zero volume. The best the demon can possibly do is to harness all the available energy of the universe to his hydraulic press. This will degrade all the world's energy to heat, effecting the heat death of the universe. Then the demon's operation can go no further. He will be out of energy.

Still the demon will have failed to banish all uncertainty in the molecule's vertical position—and he will be maximally uncertain about the state of the universe.

The surprising conclusion of information theory is that experiments done within a system can never increase total knowledge of that system at the most fundamental level. All knowledge is counterfeit. Acquisition of knowledge about one part of the world requires an equal sacrifice of knowledge about other parts. Ignorance can at most be shifted around.

The limitations of science are more extreme than the most pessimistic pre-twentieth-century philosophers would have imagined. In its most all-encompassing sense, knowledge cannot be advanced at all. Nothing—not electron microscopes, not particle accelerators, and not radio telescopes—can take us one iota closer to Laplace's ideal of knowing the details of every particle in the universe.

Such statements of Szilard's and Brillouin's theorems should not be misconstrued. Things aren't totally hopeless. The appar-

ent advancement of knowledge hinges on the distinction between information and *useful* knowledge. Useful knowledge
is information that someone actually cares about. Information
theory deals with information only as an abstract quantity.

You turn on a lamp to locate a chair in a dark room. Information theory claims that the information gained about the
chair's location is canceled out by the information lost about
the positions and velocities of the atoms in the light bulb and
surrounding air. This is entirely true, but you probably didn't
need the atomic information.

As long as it is possible to swap irrelevant information for
useful knowledge, there is no problem. Science is not and
cannot be a quest for a complete knowledge of the universe.
Rather, it is a process whereby certain information is selected
as being more relevant to human aims and understanding.

Economies are possible. Nothing in Szilard's or Brillouin's
analyses prevents us from deducing natural laws. A natural law
is an induction—a generalization from observed situations that
is assumed to hold true for unobserved situations as well. From
a small number of observations, it is possible to gain (indirect)
knowledge about what is going on throughout the universe.
This is the route physics has taken.

·4·

GLIDERS AND SPACESHIPS

Heisenberg once wrote of quanta, "I would guess that the structures with which we are confronted are beyond any objective description in imaginable terms and that they are a kind of abstract expression of the laws of nature rather than matter." This idea finds its apotheosis in Life. All Life objects are expressions of Conway's rules. Even so, there is no simple way of deducing the variety of Life phenomena from a mere statement of Life's rules. This fact likely reflects a similar difficulty of real-world physics.

The glider is not the only Life object that moves. Conway used the term "spaceship" to describe moving patterns. Spaceships are a good example of the convoluted relationship that may exist between object and physical law. As simple as the notion of a moving pattern is, there is no general procedure for finding new spaceships. Much effort has gone into finding new moving patterns.

To qualify as a spaceship, a pattern need not shift frozen from pixel to pixel. Remember how the glider moves. It evolves through four phases to move one pixel diagonally. The known spaceships are oscillators that manage to translate themselves so many pixels per cycle. A spaceship's speed is its shift (in pixels) divided by its cycles (in generations). It is far easier to discover new still lifes or oscillators than spaceships. Only a few fundamentally different types of spaceships are known, and only the glider is common.

SPACESHIPS ARE ASYMMETRIC

What must a spaceship look like? There is at least one obvious prior constraint. Spaceship motion is asymmetric. A spaceship cannot move in all directions at once; it must prefer one direction over all others. That preference cannot come from Conway's rules. A pixel's future state is determined by the *number* of neighbors but not the *position* of those neighbors. If there is any asymmetry in a Life pattern's behavior, it must come from an asymmetry in the pattern itself.

The glider is completely asymmetric. There is no way that a given glider phase can be rotated (less than a full 360 degrees) to reproduce its original orientation. Nor can its mirror image be superimposed on itself.

A spaceship doesn't have to be completely asymmetric. It can have bilateral symmetry—the symmetry of an object that is the same as its mirror image. The boat and loaf have bilateral symmetry, as does the human body or a rocket.

No spaceship can have greater-than-bilateral symmetry. Suppose it is claimed that a pattern with the symmetry of a beehive is a spaceship that moves to the right. Because of the impartiality of Conway's rules, it would have to move to the left as well. Something has to be wrong.

Life's rules also limit the speed of spaceships. Suppose a certain pattern is contained in a 100-by-100-pixel square. Everything outside the square is in the off state. No matter what the structure of the pattern, it can influence only the pixels bordering the square in the next generation. It will then be contained in a 102-by-102-pixel square. At very most, a pattern can grow one pixel per generation in any direction.

This is the maximum speed at which any form of information can be transmitted across the Life plane. It is the counterpart of the speed of light in the real world and is often called by that name. A glider is said to travel at one-fourth the speed of light diagonally.

It is easy to find patterns that grow at the speed of light for a while. Any long row or column of pixels does, creating complex video effects. But the waves of births grow two pixels shorter with each generation. Only an infinite orthogonal line will grow out in both directions forever. Conway concluded that half the speed of light (orthogonally) is the limit for a finite spaceship. At best it can advance a pixel, take a generation to fill in the leading edge, and start anew.

LIGHTWEIGHT, MIDDLEWEIGHT, AND HEAVYWEIGHT SPACESHIPS

The known spaceships fall into several natural classes. The glider, unique in many ways, forms its own class. The second class has three members. It too was discovered early in the investigation of Life. One of Conway's colleagues saw a shimmering form moving across a video display and stopped the computer before the moving form crashed into another object. The moving pattern was a "lightweight spaceship."

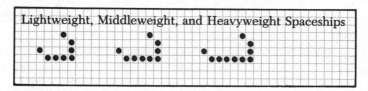

Lightweight, Middleweight, and Heavyweight Spaceships

The lightweight spaceship can be stretched into a "middleweight" or "heavyweight" spaceship. The illustration shows the three spaceships in the form that is easiest to enter into a computer (though spaceships are easily tracked on graph paper, too). This phase has a boomerang shape comparable to two of the glider phases. Unlike the glider, these spaceships travel horizontally or vertically, never diagonally. All three move toward the right when oriented as shown.

The Middleweight Spaceship in Action   Tail spark   Belly spark

Time 0   Time 1   Time 2   Time 3   Time 4

Each spaceship advances two pixels every four generations (half the speed of light). A boomerang phase is followed by a dense phase. Most of the latter dies out from overpopulation, creating an upside-down boomerang. This gives rise to an upside-down dense phase, and it restores the original boomerang —shifted two pixels.

Spaceships throw off "sparks," debris that fades instantly. The boomerang phase of all three spaceships has a single-pixel "tail spark." The middleweight spaceship also has a single-pixel "belly spark." In the heavyweight spaceship, the belly spark is two pixels. Because the sparks vanish without affecting the rest of the spaceship, they can be omitted when entering a spaceship in a computer.

On a high-speed video display, these spaceships suggest a turbine. The way the phases flip over creates a convincing illusion of 3-D rotary motion. The apparent axis of the turbine is in line with the direction of travel.

Another unexpected video effect is a "hole" near the front of the spaceship. The two dense phases each have an internal empty pixel. The corresponding pixel is also off in the boomerang phases. The hole moves at the same speed as the rest of the spaceship.

It looks as if longer spaceships can be constructed from this same design, but they can't be. The belly spark is the spoiler. A boomerang pattern one unit longer than the heavyweight spaceship creates blinkers for belly sparks. Instead of vanishing, the blinkers remain and interfere with the would-be spaceship. Ultimately, all the debris annihilates itself. Similar problems plague longer boomerangs.

GLIDER AND SPACESHIP COLLISIONS

What happens when a glider or spaceship collides with another object? The results are amazingly diverse, and most of our intuitions about real-world collisions do not translate very well. Colliding Life objects rarely bounce back (though it can happen). Sometimes they smash into many pieces. Rarely do the original objects survive. For simple, two-glider collisions, the commonest outcome is mutual annihilation.

The subject of collisions is more complicated than might be

thought, for there are usually many different ways that two Life objects can collide. Position and timing can (and almost always do) make a difference. A glider or spaceship can encounter a target object in any of the four phases with different results.

There are seventy-three distinct collisions of two gliders. A two-glider collision may be head-on or at right angles. Two gliders may form a block, blinker, beehive, pond, eater, or a single new glider. In general, two colliding gliders form an unstable mass that evolves for some time and then stabilizes. Parts of the collision mass may die out; other parts may give rise to stable objects.

In one collision that produces a block, a four-pixel flattened U shape is created at time 2. (In each diagram, time 0 is the last generation before the gliders interact.) The U shape and the three-pixel L shape are the most common predecessors to the block. Experienced Life players can recognize predecessors to common stable forms. The smallest predecessor to the pond is the four-pixel slanted Y shape that occurs at time 2 of the pond collision. Its evolution includes the characteristic chevron of time 4.

The collisions of two spaceships, or of a glider with a spaceship, are all the more numerous. Two lightweight spaceships can annihilate in a head-on collision. A slight difference in spacing between the spaceships results in an entirely different outcome: two gliders rebounding from the collision site. When the two original spaceships are positioned as shown (mirror-image boomerang phases), an even number of pixels between the spaceships will result in annihilation; an odd number of pixels' spacing will result in the two gliders.

In another collision, two middleweight spaceships can collide to form a pulsar. The spaceships first fragment and form a pair of T tetrominos. The T tetrominos split into four and then coalesce into a pulsar.

Gliders and spaceships can also collide with nonmoving objects. A glider can collide with a block and annihilate both, for instance. The eater's digestion of a glider is really a collision of a glider with an eater in which only the eater survives. The eater can also eat lightweight and middleweight spaceships. Heavyweight spaceships cause indigestion. The eater consumes the front part only, leaving a loaf.

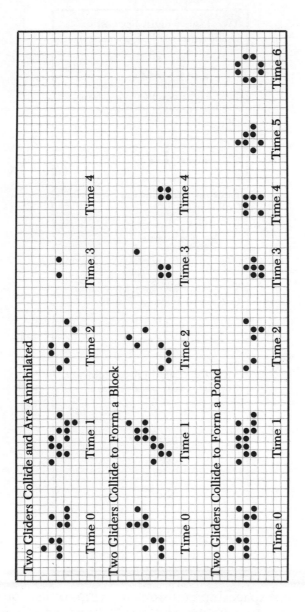

Two Gliders Collide and Are Annihilated

Time 0   Time 1   Time 2   Time 3   Time 4

Two Gliders Collide to Form a Block

Time 0   Time 1   Time 2   Time 3   Time 4

Two Gliders Collide to Form a Pond

Time 0   Time 1   Time 2   Time 3   Time 4   Time 5   Time 6

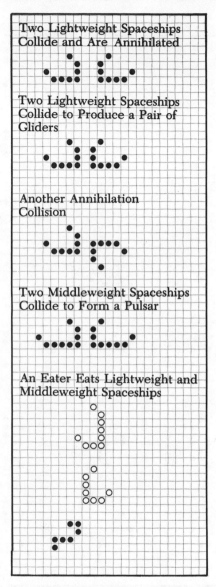

Glider and spaceship collisions are sometimes compared to the collisions of subatomic particles. There is usually more than one possible outcome of a subatomic collision. When two protons collide, the result is sometimes a neutron, a proton, and a pi meson; sometimes a kaon, a proton, and a lambda particle; and sometimes still another assortment of particles. Physicists know of no way of predicting the result of such collisions. The uncertainty principle proscribes such foreknowledge.

A common Life experiment is to position two or more gliders or spaceships on an arbitrary collision course and watch what happens. The results seem as unpredictable as those of subatomic collisions. Of course, Life is deterministic, so any precisely specified collision can have only one outcome. Nowhere does chance enter in.

One of the longstanding controversies of quantum physics is whether the uncertainly principle allows the real world to be deterministic at some basic level. "Surely God does not play dice with the universe," Einstein objected. He felt that physics must be deterministic and that the uncertainty principle merely expresses our ignorance of the subatomic world.

Einstein postulated "hidden variables." These would be unknown attributes of subatomic particles and fields. The hidden variables would determine which of the possible outcomes of a quantum collision actually occurs. If one could uncover the hidden variables, one could predict subatomic phenomena as accurately as planetary orbits.

In the 1930s, Von Neumann analyzed quantum theory from a mathematical standpoint. He showed that the known behavior of quanta was all but incompatible with hidden variables, at least of the type that most physicists were looking for. This broke the spirit of the hidden-variables contingent, save for diehards such as Einstein. Today physicists mostly favor the "Copenhagen interpretation" of quantum physics (named after Niels Bohr's Danish institute). This links reality to the act of observation—states of quanta do not even exist until they are observed.

In Life, at least, the two viewpoints need not be mutually exclusive. Two distant gliders approach, collide, and form a pond. A determinist can argue that the spacing and phase differences are the hidden variables. An equally valid "Copenhagen interpretation" of Life can maintain that collision results are tied to the act of collision. In isolation, all gliders are alike. Neither glider has some hidden attribute that means "pond."

OVERWEIGHT SPACESHIPS AND FLOTILLAS

After some thought, Conway decided that there is a way of fixing the belly-spark problem on long spaceships. The trick is to arrange for escorting spaceships to prevent the formation of sparks longer than two pixels. A group of spaceships traveling in parallel is called a "flotilla." Flotillas incorporating "over-weight spaceships" are the third basic class of spaceships. Overweight spaceships cannot travel alone, so it is the flotilla as a unit that qualifies as a spaceship.

The simplest flotillas consist of an overweight spaceship sandwiched between two regular (lightweight, middleweight, or heavyweight) spaceships. All three spaceships flip over every two generations. The escorting spaceships ensure that the margin cells of the overweight spaceship have too many neighbors to allow long belly sparks. When all three spaceships are in phase (all in the boomerang configuration simultaneously), the maximum length of the base of the overweight spaceship is twelve pixels. Then only heavyweight spaceships will work as escorts.

A slightly longer flotilla is possible if the overweight spaceship is out of phase with its escorts. The overweight spaceship must be in the dense phase while the escorts are in the boomerang configuration, and vice versa. Fourteen pixels (base of boomerang phase) is the limit.

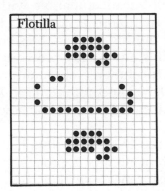

It is not possible to escort longer overweight spaceships by a train of regular spaceships on each side. Either the escorting spaceships are so close that they interact with each other or problem sparks form in the chinks between them.

Conway found a way of escorting overweight spaceships of any length. A very long overweight spaceship can be sandwiched between two overweight spaceships that are slightly shorter. The latter spaceships' outer edges can be shielded by a pair of spaceships that are slightly shorter yet. Such a flotilla assumes the shape of a double pyramid. Lightweight, middleweight, or heavyweight spaceships must form the apexes of the pyramids.

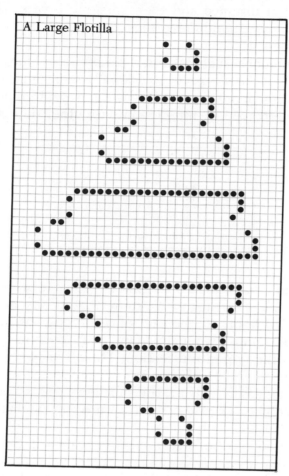

All flotillas move like a single spaceship: orthogonally, at half the speed of light. Flotillas are engineered objects—no one has ever seen one arise naturally.

SHUTTLES

It is clear that any spaceships other than the ones already described will be unnatural. Enough people have watched enough Life fields long enough to rule out the possibility of any undiscovered common spaceships.

Many failed spaceships are known. These are objects that move for a time and then run into trouble. Some patterns move a short distance and then abruptly reverse direction. Under the right conditions, they can shuttle between two points endlessly.

The commonest such pattern is simply called the "shuttle." It is a fairly natural object, occasionally seen in random Life fields. It occurs in generation 774 of the *R* pentomino, though it is soon destroyed by interaction with other debris.

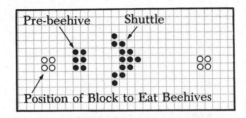

The shuttle has fifteen phases. Most look like arrowheads or rockets. It moves orthogonally, somewhat fitfully, at about 0.4 the speed of light. At the end of a fifteen-phase cycle, the shuttle splits into two parts. One part becomes a shuttle pointed in the opposite direction. The other part becomes a beehive.

Fifteen generations later, the shuttle reverses, producing a second beehive. Then the shuttle rams into its first beehive and is destroyed.

The shuttle can be preserved indefinitely if the beehives are removed. Properly positioned eaters will eat the rectangular latent beehives. Blocks (shown in diagram) can also eat the beehives. The block-beehive reaction is explosive, creating a shower of sparks and a pre-block that instantly repairs itself. The system of shuttle and blocks is an oscillator with a period of 30.

A similar shuttle can be created with a pair of "*B* heptominos." The *B* heptomino is a seven-pixel figure investigated by Conway. (The "*B*" is Conway's arbitrary label.) Like the *R* pentomino, the *B* heptomino is a small unstable object that frequently turns up in random Life fields. Left to itself, a *B* heptomino takes 148 generations to stabilize into three blocks, a ship, and two gliders.

The *B* heptomino looks like the front end of a spaceship in the dense configuration. In fact, the first few generations of the *B* heptomino include a leading edge that moves outward like the front of a spaceship, flipping over every two generations. After much experimentation, it was discovered that twin *B* heptominos can be penned between four blocks in a stable shuttle.

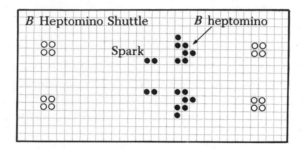

This shuttle has a period of 46. The illustration shows the phase in which the *B* heptominos occur in standard form. This phase includes two sparks that need not be entered in a computer.

The *B* heptomino shuttle is swifter, traveling at half the speed of light. As the twin *B* heptominos approach either end of their circuit, they turn around, leaving a complicated cloud of debris. Fortuitously, it is possible to position a pair of blocks just so that they annihilate the debris and repair themselves.

There is an unexpected degree of freedom to the *B* heptomino shuttle. As many as two of the blocks may be removed, as long as they aren't both from the same end. The reaction is then different, but the remaining blocks still manage to kill off the debris clouds before the *B* heptominos return.

·5·

INFORMATION AND STRUCTURE

Claude Shannon, the founder of information theory, was an engineer at Bell Laboratories. He was concerned with such problems as the transmission of television images over telephone lines. This work got him thinking about the nature of information and about how such a subjective notion might be defined.

To get anywhere, Shannon felt he would have to renounce all interpretation of information. It was just as difficult to transmit a meaningless image as a meaningful one. It seemed natural to say that they had the same information content. By the same token, Shannon was able to incorporate the notion of structure in his theory. A structured though abstract image could be distinguished from a formless image. These distinctions can be illustrated in one last version of the Maxwell's demon paradox.

AN INFORMATION DEMON

Videotape is a digital medium. It encodes all the information in a television image magnetically. Each second of action is broken down into 30 still frames, and each frame is broken down into thousands of pixels. The pixels have a certain number of states representing color and brightness. Sound tracks are digital too. It follows that there are only so many possibilities for a frame of videotape—and only so many ways of stringing frames together.

Upon hearing of this, Maxwell's demon started a new enterprise. He called it the Universal Video Library. He built a huge warehouse and filled it with videotapes. He took out ads in the show-business trade papers, to this effect:

ATTENTION, PRODUCERS: Why make TV shows and movies the old-fashioned way? Why pay millions for actors, actresses, directors, writers, set designers, location filming, special effects, editors, and sound mixers when all you really need is a videotape of the final product? Whatever you're working on, *stop production immediately.* Don't spend another penny until you come to the Universal Video Library.

The Library is guaranteed to have a completed tape of whatever you are working on—regardless of the stage of production—and I will sell you the tape for a nominal price, far less than the cost of a day of taping. How is this possible, you ask? Why, only because the Library is a Universal Video Library!

It contains a copy of *every possible* videotape 100,000 frames long (a little over 56 minutes running time). Even if two tapes differ only by the color of a certain pixel in the 59,003rd frame, the Library stocks both. Film producers, take note: The Library uses such high-resolution videotape that it may be transferred to film without any graininess. And, of course, any movie longer than 56 minutes may be found in two or more of our standard-length tapes.

You'll be amazed at the selection! Think a few Emmys or Oscars would look good on your mantle? All the future winners are somewhere on the Library's shelves. Come in and browse—I'll help you separate the wheat from the chaff. One visit to the Library, and you may never deal with a "creative" person again! All a director does is to choose one *preexisting* possibility out of a big, *finite* ensemble. You can do the same thing in the Library—cheaper. But, remember: There are only so many masterpieces, and your friends are already picking over them. Hurry in today!

THE INFORMATION DEMON REBUTTED

Maxwell's demon will find himself as incapable of sorting good videotapes from bad as he is of sorting fast molecules from slow. His downfall is again that elusive concept, information. The Universal Video Library is an update of physicist George Gamow's playful idea of a universal printing press that would print all possible lines of text. The library itself invokes no (logical) impossibilities. For that reason the demon's videotape-sorting scheme is all the more tantalizing.

The demon could certainly build a machine to generate successively all possible videotapes. The information in any videotape can be encoded as a string of digits. Each digit would stand for the state (color or degree of brightness) of a given pixel in a given frame, or for information about the sound track accompanying a given frame. Such strings of digits would be the "call numbers" of the demon's library. The machine need only run through the call numbers successively, manufacturing a tape to match each call number.

The first tape in the library would be 0000000000 . . . 00000000. (The full number would be billions of digits long.) Let 0 stand for a white pixel and also for silence in the sound track. Then 00000 . . . 00000 is a silent blank white screen for 56 minutes. As it happens, the Korean video artist Nam June Paik made a movie called *Zen for Film* that is nothing but a blank white screen. The demon's library has a hit the first time at bat—the very first tape contains a film that was shown in theaters.

The next tape is 00000 . . . 000001. Let 1 stand for a red pixel. This tape is exactly the same as 0000 . . . 0000/*Zen for Film* except for a red dot in the lower right corner of the very last frame. It will take the tape-generating machine a very, very long time to come up with any tape that doesn't start with a blank silent screen. By the same token, its plodding progress ensures that no possible tape will be missed. If there are ten states for each pixel or element of sound track, then the last tape will be 999999 . . . 999999, which might be a black screen accompanied by a monotone.

In between 0000 . . . 0000 and 9999 . . . 9999 will be a greater variety than the demon can imagine. There will be every conceivable abstract video animation. There will be talk shows and pornographic films from other planets. There will in fact be movies of everything that has ever happened or will ever happen, regardless of whether there was or will be a camera crew there to record it.

The demon will find that substantially all forms of communication and art are encompassed in the library. There will be tapes whose frames are the canvases Picasso destroyed. Sound tracks will include all possible forms of music, vocal, instrumental, or synthesized. Some tapes will be videotext, so the demon's library will not fail to include anything in any text library.

One obvious get-rich-quick scheme can be rejected immediately. Once the demon's library is complete, he can honestly claim to have tomorrow's TV newscasts of stock prices, racing results, and lottery numbers. But here the library's universality defeats him. It must also contain every possible wrong forecast. The library is useless as a crystal ball.

As long as the demon sticks to tapes that can be appreciated on purely aesthetic grounds, he seems to have a gold mine. All he has to do is sift through the shelves, throwing the junk in one pile and the good tapes in another.

Where, exactly, does the demon's scheme falter? There is both a practical problem and a more fundamental one. It is worthwhile to look at both.

There is not nearly enough videotape or warehouse space on earth to support the demon's library, of course. The number of distinct videotapes can be estimated. In effect, Shannon's information theory defines the information content of a videotape using this number.

The numbers that arise in such calculations are far, far larger than any that arise in mere physics or astronomy. It is not only convenient but imperative that exponents be used. An expression such as 10^6 means 10 multiplied by itself 6 times: $10 \times 10 \times 10 \times 10 \times 10 \times 10$, or one million. If you like, think of 10^6 as meaning the number written by putting 6 zeros after a 1.

Likewise, 10^{100} means 10 multiplied by itself 100 times, or a 1 with 100 zeros after it: 10, 000, 000, 000, 000, 000, 000, 000, 000,-000, 000, 000, 000, 000, 000, 000, 000, 000, 000, 000, 000, 000,-000,000,000,000,000,000,000,000,000,000,000,000,000. This number is called a googol. A googol is big enough that it has no direct physical meaning. It is a fantasy number that can never come up in the counting of real objects. By most estimates, the googol is trillions of times bigger than the number of elementary particles in the observable universe.

Directly below the entry for *googol* in any big dictionary is *googolplex.* A googolplex is 10 multiplied by itself a googol times. It could be written as a 1 followed by a googol zeros, were there enough paper and ink in the universe to write a googol zeros. More economically, the googolplex is expressed as a double exponent: $10^{10^{100}}$. The googolplex is all the further removed from reality than the googol.

Double exponents will be needed to express the number of tapes in the demon's library. Consider first the possibilities for a single frame of tape. Each frame specifies (in magnetic code) a still picture composed of a large number of pixels. Each pixel may be in any of a small number (usually) of states. The arithmetic is simplified if we assume ten states as above—say, white, black, and eight colors or gray tones.

The more pixels in a frame, the sharper the image. Generally a videotape image is less clear than a film image. Movie film is an analog medium. Film images are not broken into pixels but are (theoretically) continuous. The demon is justified in claiming that a digital image may be just as good, however. The eye has a limited resolving power. Beyond a certain approximate threshold of resolution, a videotape image will appear just as sharp as any image. The threshold depends on the size of the screen, the distance between the screen and the viewer, and the viewer's eyesight. Under typical viewing conditions, the threshold can be estimated as a million (10^6) pixels.

The number of possible configurations of a million-pixel screen is 10 (the number of pixel states) multiplied by itself a million times: $10^{1,000,000}$ or 10^{10^6}. You can think of this number as representing, approximately, the number of meaningfully distinct still pictures. It measures the degree of freedom in creating a picture.

Each frame is accompanied by a ⅟₃₀-second segment of sound track. The number of possibilities depends on the fidelity of the sound track. In all cases, this number is far, far smaller than the possibilities for the image. This stems from the fact that sound is one-dimensional and a TV image is two-dimensional. It will not affect matters much to assume that the image has slightly less than a million pixels and that the number of possibilities for the image plus sound track works out to exactly 10^{10^6}.

This is just one frame. Each of the demon's videotapes has

100,000 (10^5) frames. The first frame of any given tape may be any of the 10^{10^6} possible frames; there are just as many possibilities for all subsequent frames. The number of tapes in the demon's library must be 10^{10^6} multiplied by itself 10^5 times. This may be written

$$\left(10^{10^6}\right)^{10^5}$$

This can be simplified to $10^{(10^6 \times 10^5)}$ and then to $10^{10^{11}}$.

Everyone knows how big 11 is. 10^{11} is a 1 with 11 zeros after it: 100,000,000,000 or 100 billion. $10^{10^{11}}$, the number of videotapes in the demon's library, can be written as a 1 with 100 billion zeros after it. Just writing this number out would fill about 200,000 books the size of this one.

But that gives no feel for the vastness of $10^{10^{11}}$. With very large numbers, there is a tendency to confuse the length of the number's decimal representation with the number itself. 1,000,000 is much, much larger than 6, the number of zeros it takes to write it out. So it is that $10^{10^{11}}$ is fantastically bigger than the measly 100 billion zeros in its decimal representation.

$10^{10^{11}}$ is far and away bigger than the googol. On the other hand, it is nowhere near as big as a googolplex. $10^{10^{11}}$ videotapes poses a clear physical impossibility: All videotapes are made of atoms, and there are not $10^{10^{11}}$ atoms in the observable universe.

Conceding that point, let's look at the demon's more inherent difficulty. How does he go about finding the "good" videotapes?

There is no point in looking on the first shelf, the one beginning with 00000 . . . 00000/*Zen for Film.* For as far as the eye can see, that shelf must contain only tapes that are blank white except for the last frame. The call numbers will be of virtually no help in locating tapes of interest. Tape 50000 . . . 0000, which has the distinction of being in the middle of the filing order, is otherwise undistinguished—it must be the same as 0000 . . . 0000 except for something at the very beginning. Any tape with a readily describable call number (such as 33333 . . . 33333; 59595959 . . . 595959; or 01234567890123456789 . . . 0123456789) will be worthless or at best an exercise in the vein of *Zen for Film.*

The tapes with coherent images and/or sound track will have call numbers with no overall pattern. It will do the demon no good, however, to look for tapes with random-looking call numbers. Most such tapes are too random.

The demon will find that any time he picks a tape off the shelves at random, it will contain video snow and white noise —a perfectly unstructured pattern of pixels and sound track elements—throughout every frame. He will find by such a sampling technique that the vast majority—99.999999999+ percent—of his tapes contain video snow and nothing else. His job is not sorting good tapes from bad so much as *finding* good tapes. He will never find a single good tape by a random search.

The library contains three classes of tapes, which the demon might diagram as follows:

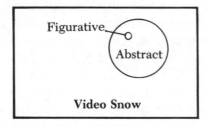

The demon is mainly interested in figurative tapes, which depict human beings and recognizable objects and settings. By no means all figurative tapes are *good* tapes. Many have banal acting, plot, and dialogue. But clearly the demon's task of finding good tapes would be far simpler if all he had to sort through were figurative tapes.

Unfortunately, the figurative tapes are far, far outnumbered by abstract tapes. Abstract tapes have structure—they aren't just video snow—and yet they don't depict anything from reality. The abstract tapes include *Zen for Film,* all nonrepresentational video art, Conway's game of Life, and all the possible TV shows from nonexistent other worlds.

The abstract tapes in turn are dwarfed by the far larger universe of video snow. The diagram cannot suggest the scope of the demon's sorting problem. The rectangle represents the $10^{10^{11}}$ videotapes. If the circle representing the number of abstract tapes were drawn to the same scale, it would be an invisible dot. Similarly, the figurative tapes are an inconceivably small minority of the abstract tapes.

The demon could draw the same diagram for his perpetual-motion machine, changing only the labels. The rectangle can represent all the possible microstates of the molecules in his chamber. Replace "video snow" with "thermodynamic equilibrium," "abstract" with "low-entropy," and "figurative" with "vacuum in lower chamber." The demon is fighting the same sort of overwhelming statistics as before.

SHANNON AND INFORMATION THEORY

Shannon noticed the similarity between thermodynamics and information. He came to formulate information in terms of the freedom of choice involved. A message specifies one possibility out of a "menu." The more items on the menu, the more information embodied in the choice. Shannon's definition purposely avoids all meaning or aesthetics that might enter into a message. He realized that these were culture-bound issues far more complex than his study could encompass.

Information and entropy fit together hand-in-glove. Von Neumann is said to have advised Shannon to adopt the term *entropy* in information theory: "No one knows what entropy is, so in a debate you will always have the advantage." Mathematically, Shannon's information and Boltzmann's entropy both take the form of the logarithm of a number of possibilities. Think of it this way: The greater the number of items on a restaurant menu, the greater your uncertainty about what the diners behind you are eating. This uncertainty is entropy. The bigger the menu, the more you must say to the waiter to make your choice clear. What you need to specify to eliminate all uncertainty is information.

Although anticipated by Szilard, information theory all but sprang up overnight with Shannon's publication of two papers in the *Bell System Technical Journal* in 1948. Besides defining information, Shannon defined the capacity of a transmission channel. This is the maximum rate at which information can be sent. Shannon proved that it is possible to transmit information reliably at this maximum rate, even when the transmission is subject to arbitrarily great amounts of garbling.

The latter result overlapped Von Neumann's theory of automatons. Von Neumann had wondered if it was possible to build a reliable machine out of unreliable components. In a sense, *all* components are unreliable because of quantum uncertainty at the atomic level. Indeed, real computer components fail to operate properly on occasion, for quantum or other reasons. Is it still possible to design computers so that they can be trusted to yield correct answers? Von Neumann concluded that it is. Computer design can overcome inherent error through redundancy. Three identical parts of the machine may calculate a certain result, each with a small chance of error. Then the three answers are compared. If the same answer occurs at least twice, it is taken to be correct. Shannon's more general result shows that the codes used for transmitting information may overcome garbling through an abstract sort of redundancy.

STRUCTURE AND MEANING

The demon's sorting of videotapes involves interpretation, Shannon's bête noire. Let the demon be replaced with an automatic device; in fact, a set of three robots. Robot A plows through the tapes, sorting into two piles: video snow and not video snow. Robot B then sorts the not-video-snow pile into an abstract pile and a figurative pile. Finally, robot C sorts the figurative pile into a good pile and a bad pile. (Robot C may also sift through the abstract pile for good and bad abstract video art.) Robot C's good pile is what the demon has been searching for all along.

Robot A looks for structure, robot B looks for meaning, and robot C embodies aesthetics. Information theory does not deny the possibility of any of these sorting operations. However, *only robot A's sorting is simple.* It can be defined objectively and is identical to Boltzmann's distinction between high-entropy and low-entropy systems. It involves only statistics.

Robots B and C must embody all the thought processes by which a human interprets and appreciates a TV show. Collectively, robots B and C are scarcely less complex than a machine that is capable of "composing" TV shows. Presumably this is possible, but it is not the easy fix the demon is looking for. If the demon knew enough about psychology and artificial intelligence to build a machine that could write scripts and then computer-animate them, he would not need the library.

The whole point of the library is to sort effortlessly, without getting bogged down in trying to codify aesthetics. The tapes that *would* be in robot C's good pile are real and on the shelves. To find them effortlessly, the demon needs information. Specifically, he needs enough information to distinguish one choice out of $10^{10^{11}}$ for each tape desired. This information is most simply specified by the call numbers of the tapes. The call numbers are long—100 billion digits—but not impossibly so.

Unfortunately, there are $10^{10^{11}}$ call numbers and $10^{10^{11}}$ videotapes. By Shannon's definition, the call numbers contain just as much information as the videotapes. The demon must supply just as much information as he gets. He will not make an information profit.

Think of how a more modest text library works. You supply a title or a call number and get a book. The book is much longer than the title or call letter. There are far more possible books of typical length than there are titles or call numbers of average length, so the book contains more information than the title or call number. You make an information profit. The demon's sorting scheme is like a library where patrons must ask for books by reciting their complete texts.

The demon will find that he can learn call numbers only the hard way. There is no way of knowing a TV show or movie's call number short of actually producing the show or movie. This is hardly the easy fix the demon is looking for, either.

The demon's library is no more useful than a simple TV set. A TV station selects one of the $10^{10^{11}}$ possible TV shows by transmitting a signal. Any of the $10^{10^{11}}$ shows can be realized by transmitting the right signal—but the station must specify the state of every last pixel in order to do so. If the demon elects to select tapes from the shelves at random, he is in the position of someone watching TV in the absence of a broadcast. The background of electromagnetic radiation still "selects" one of the $10^{10^{11}}$ possibilities and the TV set plays it. In principle this random selection could be a coherent show. In practice it is sure to be video snow and white noise.

ENTROPY, INFORMATION, AND CHANCE

Just as information is related to entropy, both are related (somewhat more loosely) to chance. Half a century before Shannon's information theory, Boltzmann wondered where the order of the world comes from. The only explanation that seemed to agree with his conception of thermodynamics was an appeal to chance.

The second law of thermodynamics is a statement of probability. The odds favor increases in entropy because there are so many more disordered states than ordered states. Nonetheless, decreases in entropy are not impossible. From time to time, there ought to be statistical fluctuations—rare instances in which entropy decreases spontaneously.

Boltzmann supposed that our universe was such a fluctuation. Beyond the known universe, he conjectured, there was a uniform sea of matter and radiation in a state of maximum entropy. Only by chance had our world's matter and radiation arranged itself into a state of low entropy. It became possible for stars to form and life to evolve. Our world's bubble of low entropy was not stable. Ultimately, it would lose its order and dissolve into the surrounding chaos.

Boltzmann anticipated the stupendous improbabilities required by his hypothesis. The chance of atoms just falling together the right way to make a universe must be inconceivably small. Boltzmann did not believe this posed a serious challenge, though. There was always the recourse to the infinity of space and time. Unlikely as the requisite fluctuation might be, even the most unlikely events must happen, given enough time (or space). Most of Boltzmann's critics objected to his belief in atoms rather than his thoughts on the origin of order.

Boltzmann was right about atoms but is almost certainly wrong about cosmic order. Boltzmann's view of creation cannot be reconciled with the known scale of the universe. There is not an infinity of past time but only about fifteen billion years. That is not enough time to make an accident of the type Boltzmann postulated plausible.

The universe is less than 10^{18} seconds old. If Maxwell's demon had been pulling tapes off the shelves of his library at random ever since the big bang, he could not have pulled much more than 10^{18} tapes. But 10^{18} is such an utterly insignificant number next to the total number of tapes—$10^{10^{11}}$—that the chance of the demon happening onto any given tape is negligible. If the demon is random in his selection, all 10^{18} tapes will almost certainly be video snow.

Suppose that every one of the 10^{81} elementary particles some physicists estimate to exist in the observable universe was replaced with a copy of the demon and his entire library, and that all the 10^{81} demons had been pulling tapes off shelves since the big bang. Together, the ensemble of demons could have collected 10^{81} times 10^{18}—or 10^{99}—tapes. 10^{99} is one-tenth of a googol and still an insignificant number next to $10^{10^{11}}$. The odds are overwhelming that the demons still would not have found a single good videotape.

Chance is unable to explain even the order present in a videotape. The universe is far, far more richly ordered than any videotape, and there an invocation of chance fails even more profoundly.

There are other problems with Boltzmann's idea. Nowhere do telescopes reveal the maximum-entropy sea Boltzmann pictured encompassing our universe. Galaxies and quasars, both low-entropy phenomena, are seen as far as our telescopes can reach.

If the universe is a thermodynamic accident, it should not be that big. Small statistical fluctuations are always more likely than big ones. The demon's library contains just one pristine copy of *Gone With the Wind,* but an astronomical number of tapes contain a few minutes of the movie or a single frame or part of a frame. In Boltzmann's sea of maximal entropy, small pockets of order—just large enough to form a single star, say —ought to far outnumber larger pockets.

The only thing special about our pocket of low entropy is that we are here. The low-entropy bubble must at any rate have been large enough for life to evolve. But most of the observed universe is surely not a prerequisite for human existence. Life on earth seems to require the sun, earth, and little else. If nothing existed beyond the outer planets' orbits in any direction, life on earth probably would not have been much different.

If the universe were a statistical fluctuation, then, it is hard to see why it is so much bigger than the solar system or at most our galaxy. A more satisfying theory of cosmic order will have to explain how there can be billions of galaxies, distant from each other and yet with the same basic structures.

·6·

UNLIMITED GROWTH

Cosmologists work downward. They try to explain the known complexity of the world using the imperfectly known fundamental physics. Life players work upward. *Only* Conway's rules are a given. It is the large-scale phenomena of the Life universe that are open questions. The game of Life is the problem of cosmic complexity in reverse.

In Martin Gardner's October 1970 *Scientific American* column, Conway offered a fifty-dollar prize to the first person to find an ever-growing Life pattern or to prove that no such pattern exists. Many readers experimented with a few patterns and gave up. It isn't hard to find patterns that grow beyond the patience of a person working with pencil and graph paper— the *R* pentomino does that. That is no guarantee that a pattern grows forever. All attempts to reason directly from Conway's rules failed too.

STRATEGIES

Many strategies for unlimited growth were tried. In its simplest form, unlimited growth might be realized by a small unstable object. It would be something like the *R* pentomino, but it would never stabilize.

The six-pixel objects were a disappointment in this respect. None comes any closer to unlimited growth than the *R* pentomino. Conway and others tracked a multitude of six-pixel objects, only to find that the most prolific patterns are those

103

that converge on the evolution of the *R* pentomino. For instance, the first generation of the *R* pentomino is a six-pixel object, and it leads to the same final constellation as the true *R* pentomino, of course.

Charles Corderman discovered a prolific seven-pixel pattern, the "acorn." The acorn is disjointed. Five of its pixels die immediately from isolation. They are replaced by six births. The acorn soon burgeons and will fill the screen of a home computer with assorted debris.

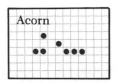

Like the *R* pentomino, the acorn grows in a haphazard fashion. Large parts of the field stabilize into still lifes and oscillators. Gliders are thrown off. Some gliders escape; others smash into other objects. A toad, an eater, and a "fleet" (a small symmetric constellation of four ships) are created and then destroyed in the acorn's evolution. As time goes on, the acorn's active regions become fewer. But occasional internal glider collisions keep the acorn percolating for thousands of generations.

The acorn does not grow forever. It stabilizes at time 5206. The acorn's vast constellation consists of 41 blinkers (including four traffic lights and a "damaged" traffic light with a blinker missing), 34 blocks, 30 beehives (including one entire honey farm and another with a beehive missing), 13 escaping gliders, 8 boats, 5 loaves, 3 ships, 2 ponds, 2 barges, and 1 mango—633 pixels in all. (The maximum pixel count is 1057, at time 4408).

Conway realized that the type of growth typified by the acorn or *R* pentomino could never settle the question of unlimited growth. Suppose a superprolific pattern is tracked for a trillion generations and is still growing. It would prove nothing. As long as the growth is chaotic, the pattern still might stabilize if only it is tracked for a few more generations. The ultimate fate of an arbitrary Life pattern is always an open question.

Only a pattern that grows in a disciplined way can demonstrate unlimited growth. Its future must be predictable like that of an oscillator or spaceship. Some Life players experimented with "chain reactions" of gliders. The idea was to crash gliders together to form an unstable mass that would emit more gliders. These secondary gliders might then collide with other gliders to form more unstable masses and more gliders in a simulation of nuclear fission. The problems of spacing and direction of gliders proved overwhelming. Other Life players tried to find examples of Conway's hypothetical glider guns and puffer trains.

THE GLIDER GUN

The shuttle is provocative—it tries to produce a beehive every fifteen generations. This seems so close to being unlimited growth. If the old beehives moved out of the way, it would be.

A group of MIT students including R. William Gosper, Jr., Robert April, Michael Beeler, Richard Howell, Rich Schroeppel, and Michael Speciner wondered if there was some way to modify the shuttle to produce gliders instead. They tried arranging two shuttles to interact periodically.

There are many ways that two shuttles can interact. They can meet at 90 degrees or 180 degrees. They can be in phase or out of phase by any number of generations up to twenty-nine. They can meet head on or a little off to the side.

Gosper's group experimented on a computer and found an arrangement of shuttles that produces a glider. Unfortunately, the glider crashed into one of the shuttles and destroyed it. After further experimentation, Gosper's group found a true glider gun.

The glider gun uses two shuttles out of phase by five generations. Every thirty generations, the shuttles interact. The result is a glider, which escapes to the southeast as shown. Two blocks placed at the far ends of the shuttles' circuits eat the beehives. (Despite the unfamiliar phases, these shuttles are the same as the type of shuttle illustrated on page 88.)

The glider gun's evolution is predictable. Every thirty generations, another glider is added to the stream. The process continues indefinitely. The glider gun is a perpetual-motion

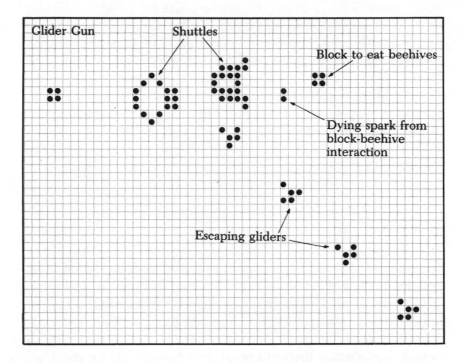

machine, an object that creates new Life "matter" without ever running down.

Gosper's group won Conway's fifty-dollar prize and soon made a remarkable discovery about the glider gun. Gliders (and nothing else) can collide to form a glider gun. Therefore, a finite number of gliders can give rise to an infinity of gliders.

The MIT group found a thirteen-glider collision that creates a gun. The gliders come from two directions (from the northeast and southeast in the diagram). The outer blocks are the easiest to create. They are formed in the two-glider collision described in Chapter Four. The two pairs of gliders so labeled make the blocks.

The shuttles are more problematic. It turns out that there is no two-glider collision that produces the shuttle. Gosper's group observed that two gliders can make a pond, however. A third glider can collide with the pond to make a ship. Finally, a fourth glider can turn the ship into a shuttle.

It follows that two groups of four gliders can make the shuttles. The shuttles must be five generations out of phase, so one

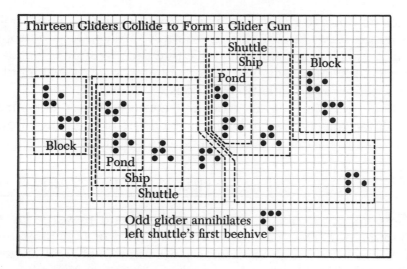

Thirteen Gliders Collide to Form a Glider Gun

Shuttle
Ship
Block
Pond
Block
Pond
Ship
Shuttle
Odd glider annihilates
left shuttle's first beehive

of the gliders that turns a ship into a shuttle must be delayed relative to its counterpart.

The thirteenth glider does not construct anything. Shortly after the left shuttle is created, it tries to produce a beehive (there being no right shuttle to react against yet). The odd glider destroys the nascent beehive, and itself in the process. If the pattern of gliders shown is time 0, the gun is assembled at time 67 and fires its first glider at time 92.

Two or more glider guns can be positioned so that their glider streams intersect. Invariably the streams react, for there isn't quite enough space for one stream's gliders to slip between the crossing stream's.

In the simplest case, the gliders from two guns can collide in just such a way as to annihilate. The gliders vanish in pairs. The system of guns and glider streams is actually an oscillator of period 30.

Usually the results are messier. The first pair of colliding gliders may form a still life. The next pair of gliders crash into the still life—which is an entirely different collision. After the first few pairs of gliders, the collision mass is almost always so active that it does not have time to stabilize before the next pair of gliders arrives. Later gliders further stir up the agitated mass.

Some collision masses seem to absorb gliders with little effect. Such masses are sometimes called "antibodies." Antibodies do not eat gliders the way eaters do, returning to their original state. Antibodies are simply active, high-entropy masses so large and unpredictable that the ingestion of a glider has no obvious effect. The glider plunges in, loses its identity as a glider, and the antibody keeps on churning.

Some collision masses grow and consume the glider guns. Collision masses are liable to shoot gliders of their own from time to time. If one of these gliders hits a gun, it will destroy it.

THE NEW GUN AND SPACESHIP FACTORIES

Gosper's group also discovered a glider gun powered by B heptominos. The "new gun" uses two pairs of B heptominos and five blocks. The heptomino pairs interact at right angles. One pair is delayed one generation relative to the other. A glider is produced every forty-six generations.

A "spaceship factory" is a gun that shoots spaceships rather than gliders. The new gun is the basis of a simple spaceship factory.

Three gliders can collide to form a middleweight spaceship. All three gliders come from different directions. It is perfectly straightforward, then, to position three new guns so that their gliders collide in the correct manner. A new spaceship is formed every forty-six generations. The diagram shows the layout schematically.

Why not use the regular glider gun? The problem is the closer spacing of the glider streams. Before the newly formed spaceship can escape, it encounters an upstream glider (the one coming from the southeast in the collision diagram). There are nonetheless ways of designing spaceship factories from regular guns. Lightweight and heavyweight spaceships can also be constructed from gliders, so it is possible (with ingenuity) to create lightweight and heavyweight spaceship factories.

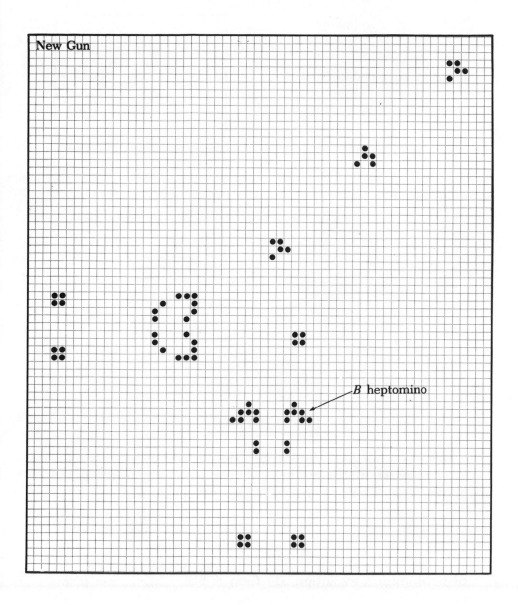

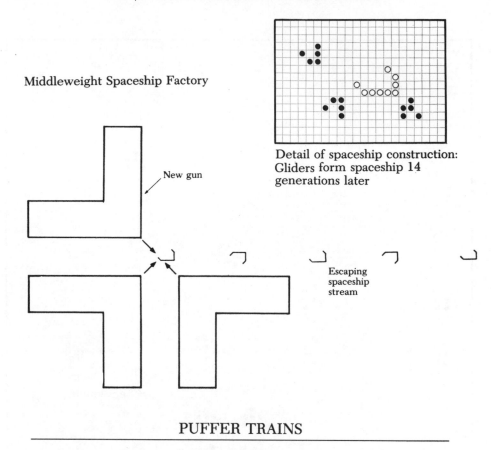

Middleweight Spaceship Factory

New gun

Detail of spaceship construction:
Gliders form spaceship 14
generations later

Escaping
spaceship
stream

PUFFER TRAINS

Conway's other hypothetical propagator was a puffer train, a
moving pattern that produces debris as it sweeps across the
plane. Many objects act almost like puffer trains.

One of the simplest is the "pi heptomino." This is the seven-
pixel form shaped like an arch or the uppercase Greek letter
pi. Generation 1 of the pi heptomino suggests a short rocket
with tail fins. This form recurs thirty generations later in the
pi heptomino's evolution. By then it has shifted nine pixels
forward. It has also created a large, roiling cloud of exhaust.

Then the cycle starts over. This time, however, there is the
exhaust cloud from the previous cycle. The exhaust interacts
with the leading edge, changing its evolution. The rocket form
does not resurface in generation 61. The would-be puffer train
is consumed by its exhaust. By generation 173, the pi heptomino crystallizes into a constellation of six blocks, five blinkers, and two ponds.

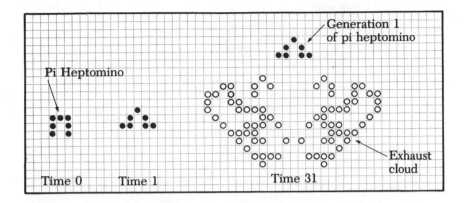

Many other puffer-train candidates share the pi heptomino's problem. Rather than produce a neat row of stable objects, candidate patterns tend to throw off wildly active exhaust. A successful puffer train must outrun its exhaust.

One format for a puffer train is a flotilla of spaceships escorting an "engine." The engine is an active object that produces "puffs of smoke"—exhaust that can freeze into stable objects. Such a puffer train travels at half the speed of light. This is the maximum speed for a finite pattern and thus the speed at which a pattern has the best chance of outracing its exhaust.

The puffer train engine must be able to keep up with its spaceship escorts, otherwise the puffer train would drift apart. The leading edge of the engine must be a sort of pseudo-spaceship. A likely candidate is the *B* heptomino.

The first successful puffer train discovered uses a *B* heptomino engine escorted by two lightweight spaceships. Its behavior is far more complex than anyone imagined—it cannot be followed fully on a home computer.

The engine is shown in the phase that precedes the true *B* heptomino as this minimizes the pixel count of the puffer train (22 pixels, leaving out the spaceship tail sparks). Every two generations, the spaceships and engine flip over and advance one pixel.

The interaction of the engine with the spaceships is precisely choreographed. The black dots represent the puffer train at time 0. The engine evolves independently of the spaceships until time 6. Then a tail spark reacts with the engine. Only the spaceships' sparks "touch" the engine—interac-

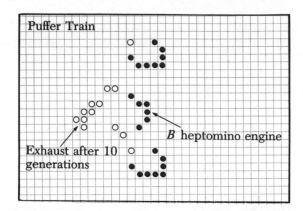

Puffer Train

Exhaust after 10 generations

B heptomino engine

tion with the spaceships proper would be ruinous. By time 10, the original configuration reappears (upside down). It has advanced five pixels and thrown off a wisp of exhaust.

The fate of the exhaust is similar to that of a collision mass caused by intersecting glider streams. Left to itself, a puff of exhaust would stabilize in due course. But the puffer train keeps ejecting new active puffs. Each successive puff reacts against a different aggregate of previous puffs. The puffer train creates an ever-lengthening plume of active exhaust. It takes a long time for the first stable, permanent objects to form in the plume.

The puffer train is one of the most striking objects to watch on a high-speed video display. Entirely by concidence, it lives up to Conway's figurative name. The evolving active regions in the plume suggest animated-cartoon clouds or puffs of smoke. Outlines of puffs undulate, grow, and ultimately disintegrate.

The convex outlines of puffs are semistable under the rules of Life. Most cells in a simple outline tend to have two or three neighbors and survive. Inward growth of a convex outline tends to be inhibited by overpopulation. Outward growth is slow.

The plume eventually gets too long for any video display. The only way to track a puffer train for long periods is to use a large computer with a printer. After a couple hundred generations, blocks, beehives, a honey farm, and loaves fleck the plume, but they are threatened by nearby active cores. As long as there are active regions in the tail, the fate of the puffer train is in question.

The long-term evolution of the puffer train is fantastic. The plume gets broader as well as longer for about a thousand generations. It comes to resemble a pyramid with the puffer train proper at the apex. Then the plume stops growing wider. It starts to look like a pencil, the pyramid being the sharpened point. Behind the pyramid is a region of constant width. This region is rich in traffic lights and honey farms (preceded by the characteristic floral effect). It is almost stabilized. Behind the region of constant width is an irregular, still-active tail region.

By time 2000, the region of constant width has stabilized completely. It becomes a symmetric frieze of stable Life objects. The active front of the plume adds on to the stable region in a predictable manner. The tail of the plume beyond the stable region is still active, however. The very end is the product of the first puffs of exhaust. Since they reacted against empty space rather than against previous puffs, their evolution is different.

These active regions still pose a potential threat to the puffer train. It is conceivable, if not exactly likely, that an active region in the unstable tail could "ignite" the plume like a fuse. In principle, an influence could propagate up the plume at the speed of light—fast enough to overtake the spaceships and *B* heptomino engine. Only by tracking the puffer train until the active tail stabilizes can this scenario be ruled out.

The active tail finally stabilizes at generation 5533. The puffer train lives up to its name and will go on creating new objects forever. The irregular end of the plume contains a pulsar and a toad (objects not found in the periodic region).

The puffer train can be modified by adding additional spaceships. A third lightweight spaceship can be positioned so that its tail sparks cause the exhaust to fade away completely. This pattern is called an "ecologist." The ecologist moves but does not produce any permanent debris. It is a spaceship, not a puffer train.

A fourth lightweight spaceship can be added for another unexpected result. The debris from the ecologist lingers for some time before vanishing. The fourth spaceship, well to the rear, can react with the last of the ecologist debris to form a glider. A new glider is created every twenty generations.

This configuration, the "space rake," is both a glider gun and a puffer train. It is trailed by an ever-lengthening front of

Time 5533

Puffer Train

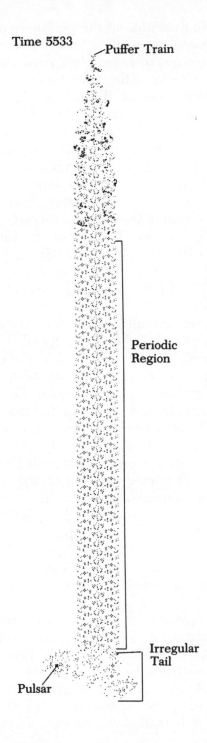

Periodic
Region

Irregular
Tail

Pulsar

gliders. Notice that the gliders move at right angles to the formation, in a glider wave rather than a stream such as the glider gun's. The space rake's gliders sweep ever greater areas of the plane as the line of gliders grows.

The space rake can be used to design other puffer trains. Take two mirror-image space rakes traveling in parallel. Position them so that their gliders collide in pairs. With due attention to phase and spacing, any desired right-angle two-glider collision is possible. It is a simple matter to produce a neat row of blocks, beehives, blinkers, ponds, or eaters as the sole puffer product. If the gliders annihilate, the system is a spaceship.

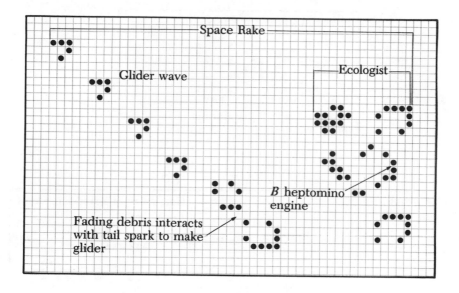

THE FLYING MACHINE

The *B* heptomino is not the only possible engine for a puffer train. The "flying machine" uses a *T* tetromino engine.

The flying machine is a spaceship with a short plume of fading exhaust like the ecologist. Two close mirror-image light-weight spaceships travel just ahead of the engine. The engine is not the *T* tetromino proper but its first generation—a seven-pixel object. Normally, the engine would evolve into a traffic light. Every four generations the spaceship tail sparks react with the engine, however. As a result the engine oscillates through twelve phases, moving forward in pace with the spaceships and throwing off puffs of exhaust that soon vanish by themselves.

The flying machine's exhaust can react with sparks from additional spaceships to manufacture stable objects. Depending on phase and spacing, a pair of heavyweight spaceships can induce a flying machine to produce gliders, blocks, or paired toads.

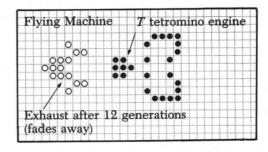

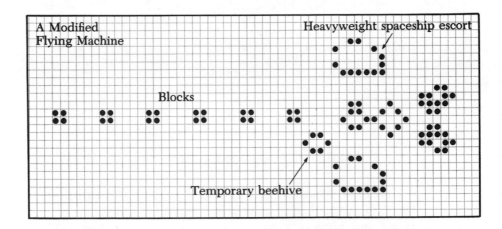

THE SWITCH ENGINE

Diagonal puffer trains are rare. All the known orthogonal puffer trains use spaceship sparks to control the engine—to keep it moving forward and prevent growth in the wrong directions. A diagonal puffer train might seem to require escorting gliders. But gliders do not throw off sparks. Any interaction between an engine and an escorting glider would have to be a direct one. The glider would be destroyed.

The known diagonal puffer trains use a bare engine. They are variations on a failed puffer train called the "switch engine."

The switch engine is a haphazard group of eight pixels. It is in three pieces: a *T* tetromino, a diagonally connected triplet, and a lone pixel.

The switch engine churns aimlessly for forty-seven generations. Then in generation 48, the original switch engine reappears. It has been mirror-reflected, rotated, and shifted orthogonally. The first two transformations cancel out forty-eight generations later, when the switch engine reappears in its original orientation. By then it has shifted eight pixels diagonally for a speed of one-twelfth the speed of light.

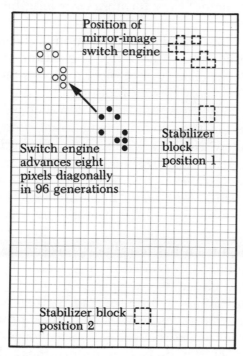

The switch engine puffs as it zig-zags, producing a large, active cloud of exhaust. The exhaust threatens to overtake the plodding engine. Eventually it does. After twelve appearances of the original orientation (time 1170), a disturbance in the exhaust overwhelms the engine.

Several ways to stabilize the switch engine's exhaust have been discovered by trial and error. One way is to place a block near the original switch engine. Puffs of exhaust will react with the block. In general, the interaction makes a big difference in the evolution of the exhaust cloud. Most of the time, the block's influence does not help much as far as the progress of the switch engine is concerned. But a few positions of the block turn out to be just right. The exhaust stabilizes and the switch engine propagates forever.

The diagram shows three ways of stabilizing the switch engine. With a block in position 1, the switch engine's plume stabilizes at time 741. It becomes a messy puffer train, producing blinkers, blocks, beehives, boats, ships, loaves, and gliders. The gliders are launched in the same direction as the switch engine's motion. Because the gliders move three times as fast as the switch engine, they form an ever-lengthening forward glider stream.

If, instead, a block is placed in position 2, the switch engine becomes a puffer train that (after the plume stabilizes) produces only blocks. The trail of blocks weaves back and forth, marking the engine's zig-zag motion.

There are many ways that two switch engines can travel in parallel. One mirror-image arrangement (see diagram) is sometimes called "Noah's ark" because of the variety of objects it produces. After stabilization, this puffer train produces blocks, blinkers, traffic lights, beehives, honey farms, loaves, gliders, boats, ships, long boats, beacons, and "block on tables" —disjointed ten-pixel still lifes. The gliders travel in three of the four diagonal directions—every direction except back into the plume. The train has four "tails," like some comets—the plume of still lifes and oscillators, the forward glider stream, and two lateral glider waves.

The switch engine is the same no matter how it is stabilized. The fact that it can form such different puffer products shows that it is the interaction of successive puffs more than just the form of the puffs that determines what will happen.

THE BREEDER

All the finite glider guns and puffer trains are alike in one way. Once stabilized, they grow at a fixed rate. At large scale, all take the form of ever-lengthening linear patterns. After a long time, a glider gun evolves into a stream of millions of gliders —plus, way at one end of the stream, the gun itself. A puffer train becomes a long, regular plume growing from one end.

Conway called his game Life because of the way the growth of patterns mimicked the growth of biological populations. Biological populations can grow without limit (given ample food, space, etc.). So can Life patterns. But in biology, even the rate of growth of a population can grow without limit. Not only are there more people in the world than there were a hundred years ago; the population is growing faster.

It is natural to wonder if Life patterns can accelerate their rate of growth. At first this sounds unlikely. It took considerable ingenuity to find constant-growth rate patterns. An accelerating growing pattern would have to find new ways to grow constantly.

Nonetheless, there are Life patterns that grow at an ever increasing rate. There are even finite patterns capable of filling all the (initially empty) Life plane.

One accelerating pattern, the "breeder," has been tracked by computer. The breeder is a large, complex pattern, too big for home computers. The breeder's premise is simple, however. Gosper's group at MIT noticed that there are puffer trains that produce gliders (the space rake, for instance) and that gliders from parallel puffer trains may collide at right angles. The glider gun can be created in a right-angle collision of thirteen gliders. The MIT group's basic idea was to design a flotilla of thirteen glider-producing puffer trains that would assemble a row of glider guns. The entire system would be a puffer train that makes glider guns.

The breeder posed several difficulties. The gliders produced by the known puffer trains were spaced too closely to construct a row of guns. The guns would have to overlap or fire through each other, both of which are impossible. What was needed was a way of eliminating alternate gliders.

The "glider train" is a complex puffer train in which four

heavyweight and two middleweight spaceships escort a pair of *B* heptomino engines. It produces two backward waves of gliders (period 32) and a pair of blocks (period 64). If desired, either or both of the lanes of blocks can be preempted by one or two additional middleweight spaceships. Optional lightweight spaceships can eliminate one or both glider waves.

The glider train's gliders are still too closely spaced. But one train's gliders can fire through another train's blocks, annihilating them. The period of the blocks is twice that of the gliders. The net result is that every other glider in the original wave is eliminated. The thinned wave is suitable for constructing shuttles: Two gliders collide to form a pond, another glider converts the pond to a ship, and a fourth glider turns the ship into a shuttle.

The breeder is a flotilla of ten painstakingly orchestrated glider trains. Blocks from two of the glider trains become the end blocks of the guns. Other blocks thin out the glider waves, which construct shuttles between the end blocks. Blocks or gliders not needed are eliminated with extra spaceships. A gun is created every sixty-four generations.

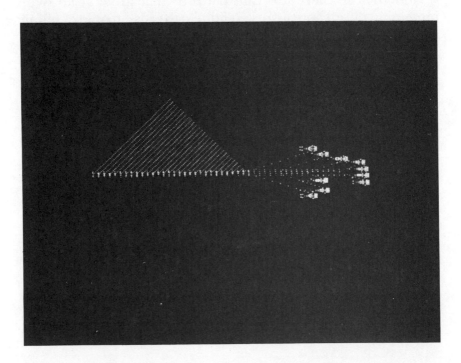

Once a gun is formed, it begins firing gliders. The more guns, the more gliders—and the faster the total growth rate. It is easy to see that the growth rate must ultimately exceed any finite value.

In time the breeder paints a vast triangle with gliders. The photograph shows a breeder as it appeared on an MIT computer screen in 1971. The individual pixels are too small to see clearly; the dots in the triangular region are gliders. The row of glider guns is visible along the base of the triangle. To the right of the base are guns in various stages of completion and the ten parallel glider trains. This photograph shows generation 3333. Thirty guns are in working order, so a new glider is created every generation. The right side of the triangle, containing each gun's first glider, expands outward at one-fourth the speed of light.

Eventually, the breeder grows to occupy a one-eighth pie slice of the empty Life plane. Eight breeders can be arranged pinwheel fashion to consume all the Life plane.

Is the breeder alive? No. The breeder is not even close to being a "living" Life pattern. Its growth is like that of a crystal. It grows, but it does not reproduce itself. It takes such an ingenious construction as the breeder to achieve crystallike growth in Life, but that does not make the breeder any more "alive."

· 7 ·

PHYSICS AS RECURSION

The first big computers at Los Alamos had a heady fascination that is all but lost today. Their strength was speed. Their weakness was that they were difficult to program. They were best suited to performing simple calculations over and over again. For the first time, mathematical experiments were practical.

Ulam introduced a deceptively simple problem to Los Alamos. It is called Ulam's problem (although he did not invent it) or the $3N+1$ problem. Think of any positive whole number. If it is even, divide it by two. If it is odd, triple it and add one. Keep applying this same rule over and over. What happens to the number?

The histories of initial values are surprisingly unpredictable. Say you choose 10. Ten is even, so you halve it and get 5. Five is odd, so triple it and add one to get 16. Sixteen is halved to 8, then 4, then 2, then 1. One is again odd, so it is tripled and added to one to get 4. The number enters the endless loop 4–2–1–4–2–1–4–2–1– . . .

Ulam found that every number he tested eventually entered the 4–2–1 cycle. Some numbers procrastinate. Twenty-seven takes 109 steps, at one point reaching a maximum value of 9232. Of course, there are larger numbers that take longer yet. But no one has ever found a number that does not enter the 4–2–1 cycle. By the same token, no one has ever been able to prove that all numbers must enter the 4–2–1 cycle. The question remains unsettled. If there are any numbers that don't enter the loop, they must be very large. All numbers up to a trillion have been tested and found to enter the loop.

RECURSION IN MATHEMATICS

"Recursive" is a mathematical term that has become popularized through such computer languages as LOGO. Simply stated, a recursive definition is a circular definition that manages to avoid paradox. When something is defined recursively, it is defined in terms of itself.

To be useful, a recursive definition must invoke a more limited version of the object being defined. What is a prime number? It's a number that can't be divided evenly by any *smaller* prime number. Notice how essential the restriction to smaller primes is. If a prime number was defined as a number that can't be divided by any other prime number, there would be no way of deciding which numbers are prime. Similarly, each value of Ulam's series is defined in terms of the previous value of the same series. Once an initial value is supplied, the series generates itself. Simple recursive series are easy to program into computers, so recursive series received much attention at Los Alamos during the war years and after.

The factorials are defined recursively. Basically, the factorial of any positive whole number is the product of all the positive whole numbers up to and including the number. The factorial of 5 (written 5!) is $1 \times 2 \times 3 \times 4 \times 5$ or 120. The precise, recursive definition goes like this: The factorial of any natural number n is $(n-1)!$ times n.

To make this definition complete, it is necessary to add that the factorial of 1 (1!) is defined to be 1. Then the factorial of any number, no matter how large, can be calculated by repeated application of the definition. What is 239! ? Well, it is 238! $\times$ 239. But what is 238! ? ... With patience, it works out that 239! is $1 \times 2 \times 3 \times \ldots \times 237 \times 238 \times 239$, as intended.

Numbers such as pi, the ratio of a circle's circumference to its diameter, are often defined as the sum of an infinite limiting series. For instance,

$$\pi = \frac{4}{1} - \frac{4}{3} + \frac{4}{5} - \frac{4}{7} + \frac{4}{9} - \frac{4}{11} + \ldots$$

This too is a recursive definition. Each subtotal obtained by adding another term is a successive approximation to the true value of pi. So the sixth-order approximation is defined in terms of the fifth-order approximation, and so forth. The series can be rewritten in typical recursive form:

$$\pi_0 = 4$$

$$\pi_n = \pi_{n-1} + (-1)^n \frac{4}{2n+1}$$

Here π_n means the sum of the first n terms. No term in the series is the true value of pi, but the bigger n is, the closer π_n is to pi.

RECURSION IN PHYSICS

One of Von Neumann's pet projects was computer weather prediction. Before computers, meteorologists plotted weather data on maps and extrapolated future changes by eye. Von Neumann felt that the computer's ability to perform repeated simple calculations would allow much more accurate prediction.

The physics of land, sea, air, and water vapor are probably known as precisely as they need to be for accurate weather prediction. Given the state of the atmosphere right now, these physical laws tell what it will be an instant from now. These same physical laws can be applied over and over to predict the weather hours or days from now—in theory, at any rate.

Unfortunately, it is rarely that easy. Errors multiply. The weather two days from now depends on the weather one day from now. Any error in the prediction for tomorrow will throw off all subsequent predictions. This is a general problem with recursive definitions. Before computers, the English mathematician William Shanks spent twenty years calculating 707 digits of pi from an infinite series. Shanks made a mistake on the 528th decimal place; all the later digits are wrong. A computer might have avoided Shanks's arithmetic error, but the errors in weather prediction are stickier. No physical measurement (as of temperature, wind velocity, pressure, etc.) can ever be a hundred percent accurate. Nor can a meteorologist place gauges at every point in the earth's atmosphere. There is always some uncertainty about the *current* weather, and the uncertainty grows the further one looks into the future.

Physics is an attempt to define reality recursively. Ideally, it is a set of rules for predicting the situation at time T_1, given the situation at time T_0. Once the situation at time T_1 is known, it can be used as a new starting point to predict the situations at times T_2, T_3, T_4, etc. Think of the intervals between successive times as arbitrarily small. Then this calculation should reveal everything that happens after time T_0.

This was Laplace's and Descartes's idea of what physics should be, and it appeals to people today. Laplace realized the impracticality of determining the details of the cosmos at any instant and the further impracticality of applying the laws of physics to this information. His point was simply that the universe is understandable, like a machine.

Laplace was inspired largely by celestial mechanics—the motions of the planets. Given a planet and its velocity at one instant, the laws of gravity predict its position and velocity an instant later. Calculus permits this recursion to be followed continuously. As it happens, the planets are all in stable orbits. The positions and velocities of the planets relative to the sun repeat with fixed cycles. Interactions between planets and countless other effects are so small as to complicate the picture only slightly.

The essential simplicity of planetary motions is evident in any attempt to predict the positions of the planets. Suppose you want to know where the planets will be 23 years and 119 days from now. One approach is to take the current positions of the planets and figure where they will be tomorrow. From that information, figure out where they will be the day after that, then the day after that, and so on. Thousands of calculations will lead to the positions 23 years and 119 days from now.

This approach embodies the mathematical spirit of recursion, but it is pointless. There is a shortcut. Without too much oversimplification, the earth will be in the same position relative to the sun a year from now, two years from now, or (exactly) 23 years from now as it is today. Draw a diagram of the planets' orbits. Figure out where the earth will be 119 days from now. It will be in essentially the same position 23 years and 119 days from now. Do the same thing with the other planets, subtracting an integral number of periods of revolution from 23 years and 119 days and figuring only the displacement due to the remainder. Just one calculation is necessary for each planet.

The brute-force calculation of daily positions fails to take advantage of the regularities in planetary motion. If the planets occupied different, unprecedented positions every day of their existence, tedious calculations would be necessary. Happily, planetary motions are as regular as clockwork. In Laplace's time, particularly, it was supposed that this simplicity had something to do with the mathematical simplicity of the laws of gravity.

But do simple physical laws guarantee simple phenomena? Physics is on the threshold of a grand unified theory whose simplicity is almost problematic: How can a world as complex as our own be that simple? Progress in fundamentally simple theories of physics has fired interest in abstract models of recursion.

Ulam and Conway were not the first to recognize the potential of recursive rules. Recursive functions of numbers have been known for centuries. The difference is conceptual. Ulam's recursive geometric objects and the game of Life are *physics*. The two dimensions of the plane suggest the three dimensions of physical space. Successive applications of the rules build a time dimension. Ulam found simple rules under which a single cell grew into fantastic networks suggesting antlers, filigrees, or Oriental designs. Physicists would like to be able to pull the complexity of our universe out of the same hat.

The underlying physics of our world is surely nothing like the rules of Ulam's games or Life, but analyses of these systems have suggested how complexity may originate from simple rules. Von Neumann showed that even the complexity of living organisms can be embodied in a relatively simple set of recursive rules. There is ample reason to believe that all the complexity of our world can be encompassed by a simple unified physical theory.

By the same token, a unified theory will not tell us everything about everything. We will not be significantly closer to Laplace's ideal of perfect knowledge of past, present, and future. Information theory limits the accuracy of any physical measurement. Quantum uncertainty places further limits. No observer can have perfect empirical knowledge of the present, much less of the distant past and future.

Abstract recursive systems suggest that it may be difficult to predict the consequences of simple rules. Conway had to discover the glider; it was not an obvious consequence of his rules. It is conceivable, then, that some of the implications of a unified physical theory could be surprising. In particular, there is no guarantee that shortcuts exist for predicting the consequences of a recursive rule.

Ulam's experiments with recursively defined geometric objects were indeed experiments. No one knew any way of predicting the pattern that would result from applying a rule a hundred times except by applying the rule a hundred times. The recursion had to be carried out step by step. So it is with arbitrary Life patterns.

Life and Ulam's games are called simulation games. They are interesting in part because they seem to simulate certain aspects of the real world. Any attempt to predict the future is itself a sort of simulation. Laplace's cosmic intelligence must represent every atom on some ethereal video screen. A computer predicting football scores, satellite altitudes, or weather must encode relevant aspects of each phenomenon somewhere in its memory. Certain shortcuts may facilitate some simulations—a planetarium model of the solar system may predict future positions of the planets, for instance—but there may be other phenomena so complex that there is no faster, simpler simulation than the phenomenon itself. It would be pointless to look for a Life pattern smaller than the R pentomino that does exactly the same thing as the R pentomino. In this sense the R pentomino may be said to be fundamentally unpredictable. If Laplace's cosmic intelligence is recast as a supercomputer, it is important to realize that no computer in the real world can possibly include all information about the real world. There may be phenomena that are forever beyond its prediction.

RECURSION AND SELF-REPRODUCTION

Part of the problem with such a supercomputer is that it must contain a simulation of itself. The simulation, in turn, must contain a simulation of itself, and so on. The nested simulations are like the reflections in barbershop mirrors. A real computer cannot possibly allow such infinite regress, since its smallest memory elements can be no smaller than quanta. The series of nested simulations must end somewhere.

There are nonetheless real physical objects that seem to contain simulations of themselves. Any organism or object that self-reproduces evidently contains enough information about its structure to produce copies of itself. This is what intrigued Von Neumann.

Von Neumann's analysis of self-reproduction is the abstract version of Watson and Crick's work. He devised a sort of universal plan for self-reproduction. Von Neumann spelled out all the crucial details, showing that neither contradiction nor external agency need be involved. Because self-reproduction can be described unambiguously, a machine can be built to reproduce, Von Neumann maintained. To put it another way, self-reproduction—the most fundamental attribute of life—can be understood as the consequence of a simplified "physics."

One of Von Neumann's first insights was that there are two kinds of self-reproduction. They are given the cryptic labels of "trivial" and "nontrivial" self-reproduction. Trivial is the simpler form and did not occupy Von Neumann much. Nontrivial is the type exemplified by living organisms and the subject of Von Neumann's extended study.

Trivial self-reproduction is best illustrated with examples. Geneticist L. S. Penrose devised several types of ingenious, trivially self-reproducing machines. One example is constructed from the two tilelike objects illustrated. Units of both types are placed against the back of a large tray held at an angle so that the units rest in a row, approximately upright.

When the tray is shaken, the units bump against each other but do not link. However, two units may be linked by hand to form a simple "machine." If two linked units are added to the tray, something unexpected happens. When the tray is shaken, new linked units form! The two-part machine reproduces itself, given the single units for food.

The two types of units can be labeled "top" and "bottom" pieces, for this describes their orientation in any possible linked pair. Both top and bottom units tend to rest flat on the

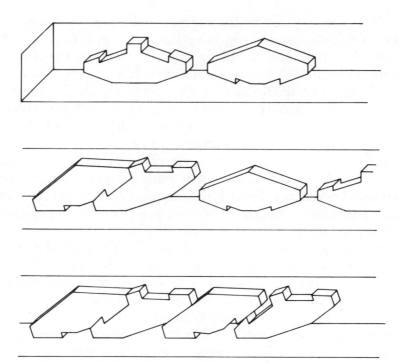

bottom of the tray. Shaking gives the units the energy to raise their centers of gravity, to rock slightly from side to side. But the shape of the units almost always prevents them from over-lapping. Even if the tray is tilted toward one end, the units tend to line up single file, touching only at the small flat sur-faces at each end.

Once two units are linked, they stay linked. The two compo-nents of the linked "machine" are at an angle to the bottom of the tray. If a tray containing a machine is tilted or shaken vigorously, wedge action forces the adjacent units to assume the same orientation. This orientation can be transmitted through a long row of units, like falling dominoes. It is not necessary that top and bottom pieces be in equal numbers or in alternating order. But wherever a top and a bottom contact correctly, they readily join up.

There are actually two ways a top and bottom can link: top on the left (illustrated) and top on the right. Whichever way the seed machine is constructed, all new machines will be the same.

There are several familiar examples of trivial self-reproduction. To qualify as self-reproduction, it is important that the food units not be able to link up without the influence of a seed machine. Penrose's tiles are something like the teeth of a zipper. In a zipper, the first pair of meshed teeth position the next pair to mesh, and so on. In that sense, zipper tooth pairs reproduce. But the pairs themselves remain linked; they cannot separate as two pairs of Penrose's tiles can.

Falling dominoes are an even simpler case of self-reproduction. The seed machine is a single falling domino, and its food is upright dominoes. There is no linkage, just a change in orientation. Similarly, the growth of crystals is a simple case of trivial self-reproduction.

Homer Jacobson of Brooklyn College built a self-reproducing toy train. Two kinds of cars, "heads" and "tails," were the units. If a head and a tail were joined, the resulting two-car train caused similar units to be assembled.

There is something unsatisfying about all these examples of self-reproduction. They are too Frankensteinian—that is to say, the units are not much simpler than the machines. Human reproduction is not taking a head and a body and joining them; it is creating a complete human being from a cell and mere molecules of nutrient.

And it is not just a question of number of parts. The problems involved in the reproduction of Penrose's tiles (getting the right angle, making sure the linkage is secure, etc.) have absolutely nothing to do with reproduction of earthly life or reproduction in general. Each case of trivial self-reproduction is a special case, with little relevance beyond itself.

Von Neumann was not interested in the details of any specific type of reproduction so much as the paradoxes that seem to be raised by reproduction in general. For this reason, he formulated the concept of nontrivial self-reproduction.

He defined nontrivial self-reproduction as follows. It requires a machine called a *universal constructor*. This is a fairly sophisticated device that can manufacture anything at all, given a blueprint of what it is to create. More exactly, it can manufacture any of a large class of possible machines that can be constructed from the raw materials available to the universal constructor.

Von Neumann's idea was to give the universal constructor a blueprint of itself. Then it would manufacture an exact replica of itself. *This* would be the interesting type of self-reproduction, for there would be no shortcuts. The universal constructor would have to manufacture each of its components from raw materials and then assemble them correctly.

Even at the time of Von Neumann's work (early 1950s), it was apparent that real life reproduces in just this way. Living cells contain universal constructors. The type of universal constructor is the same, essentially, for all life, plant and animal. Only the genetic material (the blueprint) is different. Thus an oak tree cell creates another oak tree cell. But if given different genetic material, it could, in principle, create any type of cell.

What types of raw materials should be permitted? Von Neumann didn't much care as along as they were all far simpler than the universal constructor. In his early analysis of self-reproduction, Von Neumann contemplated using various elements of 1950s technology: vacuum tubes, batteries, motors, tape, dials, photoelectric cells, and assorted structural elements. The important thing was that the self-reproducing know-how reside in the aggregate machine rather than in any of the raw materials.

THE PARADOX OF SELF-REPRODUCTION

Any analysis of nontrivial self-reproduction immediately runs into a paradox, the central problem of self-reproduction. It has already been mentioned briefly; you may have spotted it.

Given a blueprint of itself, a universal constructor constructs itself. But that isn't quite self-reproduction. It was the combination of the universal constructor *and* the blueprint that created the new universal constructor. The creation isn't quite the same as the thing that created it.

You can get rid of the blueprint—kind of—by taking it as one of the raw materials. Scatter copies of the blueprint around the universal constructor along with whatever else it uses for raw materials. In that environment, the universal constructor self-reproduces.

But that would be trivial self-reproduction once again. The

know-how of self-reproduction is surely in the blueprint as much as it is in the universal constructor. No good.

Okay, try this. Give the universal constructor a blueprint of itself *and* a blueprint of the blueprint. The constructor will churn away and produce (1) a new universal constructor with (2) a blueprint of itself. The new constructor already has its own blueprint, so it will, in turn, crank out a third universal constructor.

Then things come to a halt, for the third constructor has no blueprint to work from. Moreover, nothing has really reproduced itself. The complete system—constructor plus blueprint(s)—gets one stage simpler each generation.

Obviously, the situation is not improved by trying a blueprint, a blueprint of the blueprint, and a blueprint of the blueprint of the blueprint. It does not help to collapse these three blueprints into one: a blueprint of (a universal constructor and a blueprint of (a universal constructor and a blueprint of (a universal constructor))). This blueprint would look something like a barbershop-mirrors reflection but would be the same, as far as the constructor is concerned, as the set of three blueprints.

It might look like the only way to achieve true, nontrivial self-reproduction is with infinitely regressive blueprints—a blueprint of a blueprint of a blueprint of a blueprint of a . . . and so on, *forever.*

Self-reproduction can't be infinitely complicated. If it were, life would not be possible in our universe. You can inscribe ever smaller blueprints within blueprints only so long before you come to the atomic scale of things. Then the chain of blueprints gets lost in the grain of our universe.

So how does a universal constructor and blueprint reproduce itself exactly? This is the problem Von Neumann set out to solve.

·8·

RECURSIVE GAMES

The surprising power of recursive rules is best illustrated with a few specific cases. This chapter briefly describes three physicslike graphics games. One is an invention of Ulam's; another is a remarkable game in which all patterns reproduce; the third is a variation on Life.

ULAM'S CORAL REEFS

The simplest recursive rule allowing growth is this: Any cell touching an occupied cell experiences a birth in the next generation. Then even a single occupied cell acts as a seed for unlimited growth. All its eight neighbors become occupied in generation 1, forming a 3-by-3 square. This leads to a 5-by-5 square, then a 7-by-7 square, then a 9-by-9 square. The square expands outward in all four directions at one cell per generation.

This growth is predictable and therefore not interesting. Ulam found ways to eliminate some of the would-be occupied cells so that the pattern acquires ever richer structure. It grows in complexity as well as size.

One of Ulam's games produces an ever-growing "coral reef" from a starting pattern of a single occupied cell. The game may be played in two or three dimensions. There is no provision for deaths; once a cell is occupied, it stays occupied. To qualify for a birth, an empty cell must meet three requirements:

(1) It must border one and only one newly occupied cell. "Border" means that it must be one of the four orthogonal neighbors. Diagonal neighbors do not qualify for a birth. A newly occupied cell is one that has experienced a birth in the current generation. Only these fresh cells can create births in the next generation.

In effect, there is a growing edge of new cells that extend the pattern. The old cells are deadwood. The restriction of births to cells that are orthogonally connected to existing cells means that the entire pattern will always be orthogonally connected. Cells that border more than one newly occupied cell are disqualified, to keep the pattern from getting too dense.

(2) It must not touch an old occupied cell other than its "grandparent" cell. "Touch" is used as in Life; it cannot be any of the eight neighboring cells. The grandparent cell is the cell that created the newly occupied cell of rule 1. The motivation for this rule is to prevent branches of the pattern from growing into old branches.

(3) Take all the cells eligible for births under rules 1 and 2, and eliminate all those that would touch each other—except those that have the same parent cell. This prevents two growing edges from growing into each other.

These rules are somewhat more complicated than Life's, but they are readily programmed into a computer. Start with a single occupied cell. It produces four births (to the north, east, south, and west) at time 1.

The arms of the cross extend outward without branching at times 2 and 3. It is easy to see that the arms or "stems" must grow a cell per generation indefinitely.

At time 4, the stems grow their first side branches. The growth of these first branches is immediately choked off by proximity to complementary branches. A second octet of side branches sprouts at time 6 and is choked off at time 13. The diagram shows the reef at time 13.

There are just twelve new cells. Eight of them tip the side branches that have just been choked off by proximity to old cells (rule 2). Only the four cells at the ends of the stems can extend the pattern. Each produces a trio of new cells at time 14.

Ulam tracked the pattern over hundreds of generations. It keeps getting more byzantine, though in a way that seems

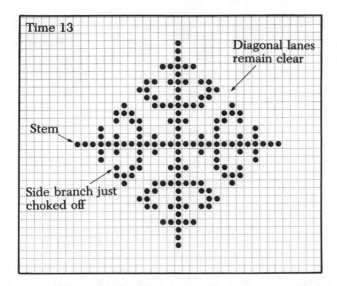

impossible to predict. Of course, it retains its original symmetry. The four diagonal lanes separating the side branches from the four stems remain clear. The density of the pattern (fraction of cells occupied) is ultimately about 44 percent.

The side branches that sprout at time 14 are still growing a hundred generations later. (The general direction of growth becomes diagonal, since no branch can cross the diagonal center lanes.) By then there are many other large side branches too. It is not known if all side branches are ultimately choked off or if some can grow forever.

Ulam wondered what happens when patterns interact. He used several disconnected cells as the starting pattern. Separate patterns can never touch, but they interpenetrate in "interference zones." Other parts of the patterns evolve normally.

Ulam used the same three rules in three dimensions. Instead of a grid, space is marked off into imaginary cubes. Each cubical cell borders six orthogonal neighbors and touches twenty-six cells all told. Ulam made a painstaking model of the thirtieth generation of a single cell with colored cubes. The three-dimensional version generates patterns with much the same labyrinth structure, but all the more intricate.

Ulam varied the rules to allow for deaths. Cells more than a certain number of generations old (usually two or three) can be erased. Then the pattern's growing fronts split apart. Parts of the pattern can move. When using the death rule, Ulam preferred to eliminate rules 2 and 3 above. With a provision for death, "fights" between two patterns may end with one pattern "killing" the other or both patterns vanishing.

Ulam found that under certain conditions, patterns could reproduce. It was a simple type of self-reproduction, more like the growth of a crystal than a living organism. Inspired by Ulam's work, Edward Fredkin devised a particularly simple game in which *all* patterns reproduce.

FREDKIN'S GAME

Created in the early 1960s, Fredkin's game uses just two states. Call them on and off. Only the four orthogonally connected neighbors count. A cell will be on in the next generation if and only if one or three (an odd number) of its four neighbors are on presently. If zero or two (an even number) of its neighbors are on, the cell will be off.

In empty regions of the plane, all cells have zero live neighbors, an even number. So regions without any live cells stay that way.

No pattern of live cells can ever die away. At the edges of the pattern at least, there must be empty cells with one live cell for a neighbor. A single cell, for instance, becomes a "tub," as all four of its neighbors have the original cell for a neighbor. In a sense, the cell has been replaced by four clones of itself in just one generation. At time 2, the four duplicates are spaced more widely, and at time 4 they are farther apart still. Every time the generation number is a power of 2, the four duplicated cells reappear. At each reappearance, they are farther apart.

Any pattern whatsoever becomes four copies. The quadrupling time is always a power of 2, but it varies with the size or complexity of the initial pattern. The "tub" that appears at time 1 of the single cell must itself be reproduced, for the rules of Fredkin's game have no way of "knowing" which was the original pattern. The pattern at time 3 is four closely packed "tubs." By time 5, the four lone cells of time 4 become four

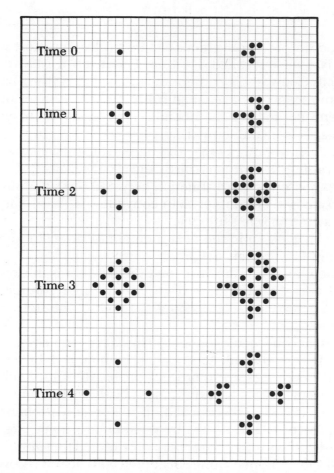

"tubs." All intermediate patterns are themselves quadrupled. Once a pattern is replicated, the set of four copies becomes the starting pattern for a new cycle of replication. One pattern becomes 4, 16, 64, 256 . . .

The diagram also shows the replication of an *R* pentomino. It takes longer to replicate because of its more complex shape. The interior of time 3 is a checkerboard pattern in which each cell has 0 or 4 neighbors and dies or remains off. This pattern, seen in all cycles of replication, puts empty space between copies.

The type of self-reproduction in Fredkin's game is trivial. Every pattern reproduces, so it is not any special feature of a pattern's organization that allows reproduction.

3–4 LIFE

Of all the simple modifications of Life's rules, the most widely played is a version called 3–4 Life. Perhaps 3–4 Life was one of the foils Conway experimented with before settling on Life. The name seems to have come from the MIT Life group. 3–4 Life seems not to have the richness of regular Life. In a sense, its appeal is in seeing how it fails.

3–4 Life is not quite what the name suggests. Rather than simply raising the survival/birth thresholds from 2–3 to 3–4, the rules obliterate the distinction between birth and survival. So 3–4 Life is even simpler than conventional Life. The rules can be stated in one sentence:

A pixel will be on in the next moment of time if and only if it has three or four on neighbors now.

The neighbors are the same eight neighbors as in Conway's game. Notice that the state of the pixel under consideration does not influence its succeeding state. Only the states of the neighbors count.

Just as no one could imagine the consequences of Life's rules, it is very difficult to say *a priori* how 3–4 Life will be different. The rules are more pro-birth: An empty cell with three *or* four neighbors experiences birth; in Life only three neighbors will do. Overpopulation is less of a problem to a growing pattern. But deaths by isolation are more likely: An occupied cell with two neighbors now dies. The rules suggest that 3–4 Life is more chaotic. Occupied cells do not receive the special consideration (via the two-for-survival clause) they do in Conway Life.

Most of what is known about the universe of 3–4 Life has been discovered by computer hacking, just as with regular Life. The 3–4 Life universe looks different. You can easily tell 3–4 Life from Life by watching a display. Life and 3–4 Life have very few stable objects in common. Of all the Life objects described in this book, only two translate into 3–4 Life: One is the block (an unusual still life in that each pixel has three neighbors); the other is the flip-flop oscillator described in Chapter Two. In it, every occupied cell dies from isolation and a mirror-image phase is created entirely by births (three neighbors).

Most stable Life objects melt away if plunged into 3–4 Life. The computer used for this book's research allows just that: A Life configuration can be used as the initial pattern in 3–4 Life (and vice versa).

Still lifes are rare in 3–4 Life. The block is the only natural still life, and it is uncommon. Notice that the two common block predecessors of regular life (the three-pixel preblock and the four-pixel form that is the lower half of a pond) both die out in 3–4 Life.

The few other known 3–4 Life still lifes are larger, engineered objects. No one has ever seen them arise naturally.

Oscillators are plentiful. Many arise naturally from simple initial patterns. The blinker does not translate, but it has counterparts in the "bleeper," the "T," the "broken T," and the

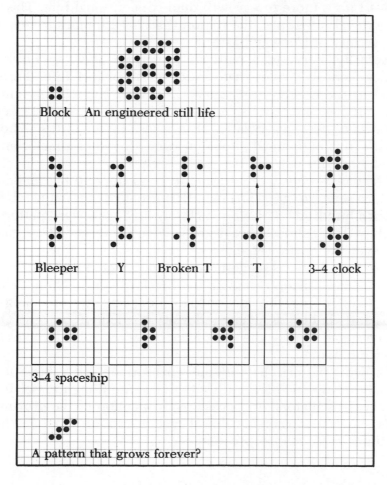

Block   An engineered still life

Bleeper   Y   Broken T   T   3–4 clock

3–4 spaceship

A pattern that grows forever?

"Y." On a high-speed display, the T seems to rotate in a plane perpendicular to the screen. All have a period of two.

An extra pixel converts a clock into a "3–4 clock," also of period two. The 3–4 clock and the T seem to be the commonest stable objects in 3–4 Life. Many other oscillators, both natural and engineered, are known. Periods of 2, 4, 6, 8, 10, and 12 are most common.

One spaceship is known. It moves horizontally or vertically at one-third the speed of light. The 3–4 spaceship's motion is crablike. The trailing end jumps forward. Then the spaceship grows backward, and then forward. The 3–4 spaceship is not as common as the glider is in regular Life. Like gliders, 3–4 spaceships can collide with themselves or other objects with unpredictable results.

3–4 Life is more pro-growth than conventional Life. This is the crucial difference, the reason that 3–4 Life fails to match the variety of Conway's Life.

Actually, very small patterns are more likely to survive under Conway's rules than under those of 3–4 Life. Three pixels is the minimum for survival in Life. Four is the minimum in 3–4 Life. This isn't so surprising—3–4 Life is stacked in favor of death by isolation. But it also favors births. When a pattern is big enough, isolation is much less of a factor and fecundity wins out.

The consequences are drastic. In 3–4 Life, *any* big random pattern seems to grow forever. All it usually takes is a patch of about a hundred on pixels in a closely knit haphazard pattern. Such patterns almost always grow like a bacteria culture, taking over the entire screen.

Growing patterns suggest crossword puzzle diagrams. Dark regions of off pixels resemble the black dividers. These holes tend to fill in from the sides once formed. The interior changes constantly but tends to have a density of about 43 percent. On a high-speed display, the short-lived holes resemble bubbles in a boiling liquid.

Growing patterns have an outward surface and a shape and usually stay in one piece rather than fragmenting into isolated still lifes, oscillators, or spaceships. Inspection of the growing edge shows that births are almost certain to outnumber deaths. So the pattern keeps spreading.

In regular Life, unlimited growth is possible but only with ingenious constructions such as the gun, stabilized switch engine, or breeder. In 3–4 Life, not only is unlimited growth easy, but there seems to be an object of just six pixels that suffices.

It is the "stairstep hexomino." It apparently becomes infinite —at least it has grown to fill the display capacity of every computer it has been run on. The growth of the hexomino follows no pattern as the glider gun does. It is more like a super-*R* pentomino. No one can prove that it must grow forever, but it seems to.

Conway was skeptical of such patterns in Life. But then there was a serious question of whether any Life object can grow forever. There is little doubt that patterns can grow forever in 3–4 Life. The stairstep hexomino is evidently the smallest.

In Life, active regions tend to stabilize. In 3–4 Life, active regions bigger than a certain minimum size tend to fill up the entire screen, engulfing anything else. The entire screen becomes a furiously churning crossword puzzle that never settles down.

·9·

BIG BANG
AND HEAT DEATH

Fifteen or twenty billion years ago, the universe was a hot, dense, formless gas. How could any set of physical laws weave the richness and complexity of our world out of chaos? Von Neumann's abstract analysis of machines begetting ever more complex machines demonstrates that complexity may increase under simple physical law—but it makes no claims about the origin of complexity in our world.

The past twenty years have yielded exciting insights on the origin of complexity. This chapter can only give a flavor of some of the current thought. It will synopsize cosmologists' reconstruction of the distant past and prediction of the distant future, a grand vision much in the spirit of Laplace.

THE PUZZLE OF GALAXY FORMATION

The universe is expanding. The distant galaxies are all moving apart from one another like the debris from a massive explosion. If the explosion is turned backward in imagination, we conclude that all the universe's matter must have been packed together at high density about fifteen or twenty billion years ago. The universe began in a big bang.

The big bang was homogeneous. This is one of the few verifiable facts about the early universe. Radio telescopes can detect microwave background radiation that was released when the

142

universe was about half a million years old. This radiation is uniform, varying by not more than 1 part in 10,000 from one part of the sky to another. It is black-body radiation, the type associated with systems in thermodynamic equilibrium.

Galaxies, their clusters, and superclusters are the most obvious features of the universe at large scale. The statistical features of the galaxy population seem constant. Astronomers don't see bigger galaxies or different types of galaxies in one part of the sky than in another. Whatever processes create galaxies, they seem to operate much the same way throughout the universe. Galaxies are held together by gravitational attraction, and all theories of galaxy formation invoke gravity.

A plausible-sounding hypothesis goes like this. Even in a very uniform gas, there will be density fluctuations. Parts of the gas will have a slightly greater density. A greater density means a greater gravitational attraction. More mass will be drawn into the denser regions, increasing the density further yet. Eventually, much of the universe's mass would be collected in these nuclei, which would be the protogalaxies.

This hypothesis has been studied intensively—and found incomplete in most analyses. Small density fluctuations would occur, and they would exert increased gravitational force. But the early universe was expanding rapidly. The gravitational effects would be too small to overcome the general expansion, even in the biggest density fluctuations that could be expected. The expansion would continue unhindered.

A more sophisticated hypothesis credits turbulence for the origin of galaxies. It can be argued that the big bang must have been subject to eddies, shock waves, and other turbulent phenomena of fluids. The turbulence could have clumped mass enough to get the ball rolling for galaxy formation. Here too, detailed studies show the effects of turbulence to be insufficient. Something else must have been at work.

THE FOUR FORCES

Cosmology and particle physics are intertwined. Increasingly, cosmologists believe that large-scale phenomena such as galaxies are explainable as implications of subatomic forces. Physics knows of four fundamental forces. Unified theories attempt to show that some (or all) of these forces are really different aspects of the same thing.

The weakest force is gravity, yet because all mass is subject to gravity, it is the force that rules the cosmos at large scale. Gravity holds the earth, the solar system, and the galaxies together. It is slowing the expansion of the universe.

The electromagnetic force is the other familiar one. It encompasses all electric and magnetic phenomena. Electromagnetic force holds atoms and molecules together. The force you feel when you press against a wall is the electromagnetic repulsion of electrons in your hand and the wall. Like gravity, electromagnetic force operates over cosmic distances. But there are positive and negative electric charges, usually present in equal numbers, so the overall effect is nil. For this reason, electromagnetic force is unimportant at the astronomical scale.

The other two forces operate only over distances comparable to the width of an atomic nucleus or smaller. They are called the weak force and the strong force. The weak force is intermediate in strength between gravity and electromagnetism. It governs certain types of radioactive decay. The strong force is the strongest known. In its pure form (the color force) it binds quarks into protons and neutrons so tenaciously that no free quark has ever been isolated. The strong force also holds protons and neutrons together in nuclei.

Forces that seem quite different may be fundamentally identical. Maxwell showed the underlying identity of electricity and magnetism. Newton demonstrated that the force that draws an apple to the earth also governs the motion of the planets. It was Einstein's conviction that all physical forces could be shown to be one. Einstein's middle and later years were spent in a fruitless search for a "unified field theory."

Later physicists have succeeded, partially, at Einstein's plan of unification. In 1967, Steven Weinberg and Abdus Salam laid the groundwork for a unification of the electromagnetic and weak forces. Experiment has since confirmed that the two forces can be viewed as different aspects of a single "electroweak" force. The grand unified theories that have occupied physicists since then unite the strong, electromagnetic, and weak forces. Debate continues over which grand unified theory is best, but there is strong conviction that the strong force is fundamentally identical with the electroweak force. Only gravity remains to be unified. Theories that attempt to unify gravity with the other forces, super-grand unified theories, have not progressed beyond outlines.

The electromagnetic, weak, and strong forces seem distinct enough. Nothing in any grand unified theory changes that. When physicists say that these forces have been unified, they mean only that there are situations in which it makes no sense to distinguish between the three forces. When the energy of interacting particles is high enough, the forces are the same.

A tray of pennies furnishes an analogy. Normally, there seem to be two types of pennies, heads and tails. Move the tray gently and the pennies will slide around. The numbers of heads and tails will not change. But vigorous shaking will toss the pennies into the air. Some pennies that start out being heads are likely to land tails, and vice versa. Whenever flipping over is a possibility, it makes no sense to distinguish between heads and tails. There is really only one kind of coin.

Similarly, the electromagnetic and weak forces become indistinguishable when particles collide vigorously enough. The minimum energy is called the unification energy. It can be expressed as a temperature—the temperature at which the average thermal collision contains the unification energy. There is another unification energy, much higher, at which the strong force merges with the electroweak. At a still higher energy, gravity may unite with the other forces.

Not only forces but particles become indistinguishable at high energies. At energies above that of the grand unification (of the strong and electroweak forces), there are really only two types of particles: bosons and fermions. Electrons and quarks are two examples of fermions. They would be indistinguishable at energies above the grand unification. The photon is a boson; it would still be distinct. Even this distinction may be erased at the super-grand unification energy. Above this energy, there would be but one type of particle and one type of force, a spare simplicity worthy of Conway.

Cosmologists believe that all these unifications were played out (in reverse order) in the first split second of the universe's history. Call time 0 the (extrapolated) instant at which all the universe's substance was compressed to infinite density and temperature. The closer the time to time 0, the higher the temperature and density must have been. Some or all the four forces were unified in the universe's first moments. The successive deunifications in the first split second had important consequences billions of years later—possibly including the formation of the galaxies.

THE COSMIC TIME SCALE

Since the 1960s, cosmologists have been able to create a remarkably complete picture of the early evolution of the universe. Temperature, pressure, density, and types of particles present have been reconstructed far back into the universe's first second of existence. A more recent step is the prediction of the long-term future of the universe. In 1979, Freeman J. Dyson examined the far future in an influential paper in *Reviews of Modern Physics.* Though subject to many uncertainties, the construction of a cosmological time scale is one of the major intellectual achievements of this century. It is the first earnest attempt to plumb the vastness of physical time. It is creating a new Copernican revolution, one just starting to enter the public consciousness.

The history of cosmology has been one of repeated diminution of the human realm. Copernicus argued that the earth is not the center of the universe. Later generations showed that the sun is not the center, either, but is a typical star out of a hundred billion or more in our galaxy. Cosmologists of this century have found that our galaxy is itself an unexceptional object out of billions in the observable universe. Cosmologists are now realizing that humanity's realm is as dwarfed in time as it is in space. The life-span of a human being, a human society, the earth, or the very universe as we know is minuscule on the time scale of cosmic history.

Cosmic history is forever slowing down. Imagine making a time line of cosmic evolution. Let each inch represent a fixed interval of time. Plot and label all important milestones: the formation of galaxies, atoms, and stars; the separation of the

fundamental forces from one another; the creation of fundamental particles; etc. The surprising thing about such a time
line is that the milestones are concentrated strongly at the very
beginning. Many important events occurred in the first split
second after time 0. Events of truly universal significance become rarer as time goes on. Their rate continues to decrease
through cosmological predictions of the future. Cosmic history
has not stopped—there are radical changes to come—but its
pace will attenuate even further. It is difficult to get a grasp on
cosmic history when lots of things happen in the first billionth
of a second (as they did) and nothing at all happens for quintil-
lions of years later on (as will be the case in the far future).

If you had a time machine and a movie camera, you could
zip back to the big bang and make a movie of cosmic evolution.
The time machine would need a fast-forward setting so that
you could travel through time at an accelerated rate while
filming. That way, billions of years of cosmic evolution could
be compressed into a reasonable amount of film.

Say the fast-forward setting is such that a billion years of
universe time passes per minute of your time. Then you would
reach the current epoch about fifteen or twenty minutes into
the movie. But wait—important things happened in the first
minute of universe time. All this action has been lost in the
movie.

If instead you decide not to use the time machine's fast-
forward feature, then the first minute of the universe is in the
first minute of the movie. But the movie would have to be
billions of years long to follow the action up to the present
time. Furthermore, most of the important events in the first
minute of the universe actually take place in the very first split
second. That's still too fast for the movie to capture.

No matter how the time machine's speed is set (even in a
slow-forward setting), the action is too fast at the beginning
and/or the movie is too boring at the end and impracticably
long. It looks as if there is no way to capture the flow of cosmic
history. Granted, the movie is a pretense, but understanding
of a process is easier with a paced flow of connected events.

The time machine needs a special control allowing it to
travel through time at a rate that increases exponentially.
Then the time machine can speed up as the pace of history

slows. Any given minute of the movie will contain a comparable amount of action. The exponential control allows unlimited compression of action.

The time machine is a way of thinking about the exponential notation cosmologists use. It becomes reasonable to measure time intervals in orders of magnitude (factors of ten in the age of the universe) rather than in absolute intervals. Think of the order of magnitude as a unit of time measured by clocks inside the time machine; it is however long it takes the age of the universe to increase by a factor of ten (outside the time machine).

Much took place in the first second, so cosmologists use negative exponents. 10^{-6} means $\frac{1}{10^6}$, or one one-millionth. The interval between 10^{-20} and 10^{-10} seconds after time 0—which is not quite $1/10,000,000,000$ second—is ten orders of magnitude. In terms of cosmic history, this interval is perhaps comparable to the interval between 1 year and 10 billion years. The accompanying time line is marked off by orders of magnitude. Times are expressed as an age in seconds after a hypothetical time 0.

It is impossible to show time 0 because there is no way to express zero as a base-10 exponent. But there is a logical point to start a cosmological time scale. It is approximately 10^{-43} seconds $(1/10,000,000,000,000,000,000,000,000,000,000,000,-000,000,000$ of a second) after the very beginning. This tiny time interval is called the Planck time, after German physicist Max Planck. Its significance here is that current physical theory is not able to say what happened before the universe was a Planck time old.

Before 10^{-43} seconds, the temperature must have been so high that gravitational effects must have been important even at the quantum scale. In normal particle collisions, the standard subatomic forces prevent particles from approaching too closely. The more vigorous the collision, the closer the approach. Gravity increases as the distance between particles decreases. In a sufficiently energetic collision, gravity should be important even between two elementary particles.

Many physicists suspect that all four forces were indistinguishable before the Planck time. The time before 10^{-43} seconds is sometimes called the super-grand unified theory

Cosmic Time Scale

Age of universe in seconds

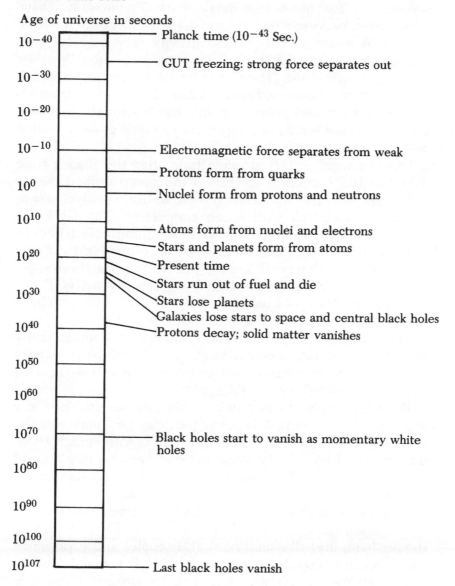

10^{-40}	Planck time (10^{-43} Sec.)
10^{-30}	GUT freezing: strong force separates out
10^{-20}	
10^{-10}	Electromagnetic force separates from weak
10^{0}	Protons form from quarks
	Nuclei form from protons and neutrons
10^{10}	
10^{20}	Atoms form from nuclei and electrons
	Stars and planets form from atoms
	Present time
10^{30}	Stars run out of fuel and die
	Stars lose planets
10^{40}	Galaxies lose stars to space and central black holes
	Protons decay; solid matter vanishes
10^{50}	
10^{60}	
10^{70}	Black holes start to vanish as momentary white holes
10^{80}	
10^{90}	
10^{100}	
10^{107}	Last black holes vanish

150 | THE RECURSIVE UNIVERSE

(SGUT) era, for it will take a super-grand unified theory to account for the physics going on then. Presumably, there would have been just one type of particle in the SGUT era. John A. Wheeler speculates that gravity may have been so strong as to change the structure of time and space. Wheeler envisions a spacetime "foam" before 10^{-43} seconds in which the curvature of time and space changed chaotically from instant to instant and point to point. But in the absence of a super-grand unified theory, physicists can only guess at events before 10^{-43} seconds.

The first eight orders of magnitude after the Planck time (10^{-43} to 10^{-35} seconds) are called the grand unified theory (GUT) era. It is the first phase of cosmic history about which physicists can speak with much confidence. The GUT era lasted less than a hundred billionth of a trillionth of a trillionth of a second, but given the compression of cosmic history, it was a momentous split second. During the GUT era, all three subatomic forces were unified (though distinct from gravity). There were just two types of particles, bosons and fermions. There was no structure above the quantum level.

As the universe expanded, the temperature decreased. By about 10^{-35} seconds, it cooled to the grand unification energy —the strong force became distinct from the electroweak. This milestone is called the GUT freezing.

The GUT freezing was crucial to the later evolution of the universe. It created new types of particles (or, more exactly, particles that had been interchangeable before ceased to be interchangeable). The universe inflated very rapidly around the time of the GUT freezing, far faster than subsequent expansion.

Grand unified theories predict that the strong force would not separate out everywhere simultaneously. Certain parts of the big-bang fireball would be slightly cooler and experience GUT freezing slightly sooner. The GUT freezing is such a radical change in physics that theories predict some side effects.

Think of a line of standing dominoes miles long. An earthquake jars the dominoes. At hundreds of places in the line, the dominoes start falling in familiar chain reactions. Some of the chain reactions go in one direction, others in the opposite direction. A given chain reaction propagates until one of two things happens. Either (1) it meets up with a chain reaction moving in the opposite direction; or (2) it encounters the fallen dominoes left by another chain reaction moving in the same direction. In both cases, the chain reaction dies. After all the dominoes have fallen, the line is inspected. Every event of type 1 leaves a permanent record. Wherever (1) occurred, the orientation of the fallen dominoes would switch abruptly. It might leave a lone domino standing, supported by dominoes leaning against it from either side—a "defect" in the line.

GUT freezing would start from cooler regions in the fireball and expand outward from each such seed to fill all of space. There are various ways that waves of GUT freezing could meet and create "defects." As in a line of dominoes, the defects would be long-lived or permanent. There is no simple way of rationalizing the properties of GUT freezing defects, but theory tells what they would be like. One type, "point defects," would have strong monopole magnetic fields and may be floating around space today. Another type, "string defects," would exert an intense gravitational field—a million tons' worth for a defect of atomic dimensions.

A defect is a form of structure that arises where none existed. A line of standing dominoes is symmetric and uniform. Nothing distinguishes any domino or position in the line from any other. After the earthquake, certain positions—the defects—are special. The line of dominoes has become more complex in the sense that it takes more information to describe it. The GUT defects (and other instances of "spontaneous symmetry-breaking") hint at the origin of cosmic complexity.

By one hypothesis, these massive GUT defects became the gravitational nuclei for galaxy formation. This idea seems to require that point defects—magnetic monopoles—be common in the universe today. Recent searches for these magnetic monopoles have found little evidence for their existence.

The "inflationary model" of the early universe explains the lack of observed monopoles as a consequence of a rapid expansion of the universe after the GUT freezing. In this model, much of the universe's energy was in the Higgs field, a theoretical field mandated by grand unified physics. As the energy of the Higgs field was released, the universe would have expanded explosively by a factor of 10^{50} or more. Since the monopoles were created before the inflation, the density of monopoles in space would be 10^{50} times less than that expected had the inflationary episode not occurred.

The Higgs field, like all quantum fields, is subject to fluctuations demanded by the uncertainty principle. The strength of the field varies from point to point in space at very small scale. The rapid inflation of the universe would have blown such quantum fluctuations up to cosmic size. The matter and energy released from the Higgs field should share the same pattern of irregularities. Theory predicts that these density fluctuations would be of all sizes. Such a spectrum of density fluctuations could provide the nuclei for galaxy formation.

The GUT freezing marked the beginning of the electroweak era. From 10^{-35} to 10^{-10} seconds, only the electromagnetic and weak forces were unified. Quarks (now distinguishable from electrons and other fermions) existed in a free state. Certain types of particle decay early in the electroweak era may have created a slight preponderance of matter over antimatter particles. Later on (when the universe was about a second old, give or take a few orders of magnitude) the antimatter annihilated. Our world is built of the residue of matter left behind from matter-antimatter annihilation.

At about 10^{-10} seconds, the weak force separated from the electromagnetic. From that point on, the physics was much as it is today: the same four forces and the same variety of fundamental particles.

Several orders of magnitude later, the fundamental particles started assembling into composite systems. The quarks joined together into protons and other particles around 10^{-4} seconds. At about 10^2 seconds (a few minutes after the beginning), some protons joined with neutrons to form simple deuterium and helium nuclei. At 10^{13} seconds (about half a million years), the universe became cool enough for atoms to form.

Before the first atoms, the universe was opaque. The hot gas of nuclei and electrons absorbed and reemitted radiation constantly. An intense radiation pressure prevented matter from clumping. But a gas made of atoms (such as air or the hydrogen-helium mixture of the universe after 10^{13} seconds) is transparent. Photons could travel long distances without interacting. After 10^{13} seconds, radiation and matter went their separate ways.

Radiation from this parting is in fact the microwave background detected by radio astronomers today. Its uniformity attests to the lack of structure in the universe at the time it was released. Once the universe became transparent, radiation pressure was less significant. It became possible for matter to collapse under gravity.

The first prerequisite for galaxy formation is local regions of increased density. The density fluctuations expected from inflationary cosmologies would be of all sizes, from microscopic to astronomical. Theories of cosmic structure must explain why only fluctuations of the proper size survived.

Many cosmologists, notably Joseph Silk and Yakov Zel'-dovich, believe that neutrinos (neutral particles believed to have a slight mass) played an important role. The universe is bathed in neutrinos; however, they interact so rarely with matter or radiation that they are hard to detect. Computer simulations indicate that the flux of neutrinos would have tended to smooth out all small density perturbations in the early universe. But if neutrinos do have mass, they would eventually slow down and be captured by the largest perturbations. Computer models predict that the neutrinos would collect into huge fibers. The fibers, much larger than galaxies, would surround voids in which few neutrinos would be found.

Neutrinos are invisible. Galaxies can be mapped, and their clusters do tend to form filamentary structures surrounding voids. The sizes of the galactic fibers are the same order of magnitude (about 100 million light-years across) as estimated for the neutrino fibers. Clusters of galaxies seem to have formed in gravitational lattices created by the neutrinos. No existing theory of galaxy formation explains everything, but there is better reason than ever to believe that the explanation for galaxies lies in fundamental particles and forces rather than in some special initial condition.

Condensing gas in galaxies eventually formed stars. By then the universe had cooled to far below the temperature necessary to spark hydrogen fusion. Contraction of gas clouds to stars reversed the general cooling trend locally. As stars became compact enough, fusion started.

The life-spans of stars vary greatly. Very large, bright stars burn their fuel the quickest. They may last a few tens of millions of years. The sun (and its planets) have been around for 4.6 billion years and are nearing the midpoint in their life-span. The first stars may have started shining 10 to 100 million years after the big bang (10^{15} seconds later). Many generations of short-lived stars have been born and died since. The present time is about 15 or 20 billion years (less than 10^{18} seconds) after time 0.

As much as the evolution of life looms large in our perception of cosmic history, it gets short shrift on the cosmic time scale. Life is a fleeting phenomenon in the universe.

An obvious prerequisite for life is planets. Perhaps life is possible on planets very different from the earth; perhaps not, and only the rare earthlike planet can support life. In any case, it seems unlikely that any kind of life can evolve in gas clouds, stars, or any nonplanetary environment.

Planets, particularly earthlike planets, are made out of different chemical elements than the universe as a whole. Most of the universe is hydrogen and helium, in proportions that probably haven't changed much since the first few minutes after the big bang. Small, rocky planets such as the earth are made largely of heavier elements such as iron, nickel, oxygen, and silicon. Large planets such as Jupiter approximate the primordial hydrogen-helium mixture but probably have cores of heavier materials as well.

The universe's heavy elements have been manufactured in stars. The fusion process that converts hydrogen to helium (the main reaction fueling stars) can continue further to produce carbon, oxygen, silicon, iron, and other heavy elements. At the end of its lifetime, a star contains much of these elements.

For most stars, this is a dead end. The heavy-element "ash" is trapped in the dying star. But very massive stars ultimately explode as supernovas. A supernova ejects much of its material violently into space. This material, rich in heavy elements, is available for the formation of new stars.

Our solar system is only about one-quarter the age of the universe. In the ten to fifteen billion years that preceded our solar system, hundreds of generations of short-lived, massive stars must have formed and produced heavy elements. Evidently, the first generation of stars did not have rocky planets. The percentage of heavy elements in forming star systems has probably increased steadily. Earthlike planets have become more and more common.

No one knows when the first rocky planets appeared. As a very rough guess, perhaps it was when the universe was about one-tenth of its present age. We likewise know few details of the early evolution of life. All the advanced forms of life— meaning by that forms as complex as fish, insects, or land plants, say—appeared in the last 10 percent or so of the earth's history. But there is fossil evidence for algaelike plants over about two-thirds of the earth's history. In any case, it seems safe to assume that the universe was lifeless for at least a few billion years after the big bang. Of all the order-of-magnitude jumps from the Planck time to the present, only the most recent has been suitable for life.

THE COSMOLOGICAL FUTURE

Has the universe settled into a final state? Or will it go on changing forever?

Cosmologists do not aspire to the atom-by-atom knowledge of Laplace's cosmic intelligence. Instead, they assume that many future events will be shaped more by the generalities of physical law than by the unknowable details of a cosmic microstate. They conclude that the universe is still evolving.

Predicting the cosmological future is tricky. The universe is no longer in thermodynamic equilibrium, as it was in most of the order-of-magnitude jumps from the Planck time to the present. No longer can the unknowable microstate be assumed random and therefore basically simple. Cosmologists must deal with the fates of galaxies, stars, planets, and dust grains as well as quanta.

Everything will come to an end if the universe contracts under its own gravity back to infinite density. The best current evidence makes this doubtful, though. Many predictions hinge on unresolved issues of physics. The long, long-term future must be determined by processes so slow that they have barely had effect so far. Such processes are difficult to measure in a laboratory. Many slow processes will work concurrently. It is easy enough to predict the result of one process, but that result may be preempted by a different process.

One way in which the universe is plainly changing is through the birth and death of stars. The sun will swell into a red giant star in about six billion years. It will vaporize the inner planets, probably including the earth, and then contract to a white dwarf star.

The death of one star does not make much of a difference at the cosmic scale, but stars are dying in galaxies throughout the universe.

The largest stars explode as supernovas and enrich the interstellar gases. There is still much diffuse hydrogen that has never condensed into stars. Cosmologists believe that stars will continue to form to replace dying stars for a long time.

Eventually, however, the available fuel will get used up. Estimates of the amount of hydrogen in galactic halos and the rate of stellar burning place this cosmic fuel crisis at about 100 trillion years from now (about 10^{21} seconds).

That is a time so far in the future that it dwarfs the current age of the universe. There is still plenty of fuel left for thousands of generations of solar-type stars. A hundred trillion years is such a long time that it makes little difference whether it is a hundred trillion years from now or a hundred trillion years from the big bang—especially as the 100 trillion years is just an order-of-magnitude estimate.

So by the time our time machine reaches 10^{21} seconds, the stars are dying or dead. The galaxies are still there and still contain most of their original mass, but they have become dark.

Life on earth requires sunlight. There is strong reason for believing, then, that life will not be possible after all the stars die out. If that is true, then life is possible only when the universe's age is in the range of about 10^{17} to 10^{21} seconds. That is just four or five order-of magnitude jumps.

The prospects for life seem to get even worse. The sun and all the stars in a galaxy are constantly moving. Different stars move at different speeds and in different directions.

Any star regularly passes, or is passed by, other stars. Sometimes the closest approach is many light-years; sometimes it is much closer. Eventually, every star must approach so close to another star that the orbits of its planets (and the other stars' planets, if any) are disrupted. A planet could gain enough energy from an encounter to be thrown out of orbit entirely. Ultimately, all stars must tend to lose their planets to interstellar space.

The process takes a long time. It has hardly happened at all so far. To estimate how long it takes, on the average, for a star to lose its planets, you need know only a few basic statistics: a typical size for planetary orbits, a typical speed for stars, and the average density of stars in space. All are known approximately.

Think of a star's train of planets as sweeping a certain volume of space every year. This volume is a cylinder. Its diameter is the diameter of the outer planet's orbit. Its height is the distance the star travels in a year.

The star must sweep out a large enough volume so that, on the average, one would expect there to be 1 star in that volume. It turns out that there is about 1 star for every 35 cubic light-years in our galaxy. This appears to be a typical density, so the star must sweep out a volume of about 35 cubic light-years for one close approach.

If the diameter of the outer planet's orbit is about 200 million kilometers and if the star moves through the galaxy at about 50 kilometers per second, then there ought to be a close approach about every quadrillion years. As a guess, it may take several dozen such approaches to knock all the planets out of orbit. This would occur by 10^{16} or 10^{17} years (10^{24} seconds).

A similar but slower process applies to galaxies. Stars are themselves in a sort of orbit about the center of our galaxy. Galaxies interact much more readily than stars do. The average spacing between galaxies is only about an order of magnitude greater than their diameters. Galaxies not only approach each other, but often collide.

When galaxies collide, their stars do not smash into each other. Because the stars are so far apart, they pass through the other galaxy's stars. The main effect on stars is gravitational. The presence of the other galaxy disrupts stellar motion. Some stars gain energy; some lose energy. Some gain enough energy to be thrown clear of the galaxy altogether. Others lose so much energy that they spiral inward to the galactic center.

Periodic collision may be largely responsible for the spiral structure of galaxies such as our own. But as it continues into the indefinite future, it will have more drastic effects. Ultimately, practically all of a galaxy's stars must either escape into intergalactic space or sink to the galactic center. (The same holds for the galaxy's planets, which would by then be mostly detached from their stars.)

Furthermore, the stars that fall in toward the center of the galaxy should form a supermassive black hole (a region with such intense gravity that nothing can ever escape from it). Actually, many astronomers suspect that there is a large black hole (or holes) at our galaxy's center right now. If stars fall into it over the eons, it will grow steadily larger.

It can be estimated that this process of galactic depletion will take on the order of 10^{18} years—10^{25} or 10^{26} seconds. By then:

- all the stars will have died;
- many planets will have been incinerated as their stars bloated just before running out of fuel;
- the other planets will have cooled off to a temperature near absolute zero and subsequently been thrown out of orbit;
- many dead stars and planets will have been thrown into intergalactic space;
- and other stars and planets will have fallen into black holes.

So far all these predictions are based on known, observable celestial mechanics and astrophysics. We are extrapolating familiar processes only.

Dyson and other cosmologists have tried to extend the cosmic time scale much further. The future beyond the events already discussed is only an object of speculation. As we go to longer periods of time, it is possible that the cumulative effect of processes currently unsuspected could be important.

One of the biggest questions affecting the far future is proton decay. To all intents and purposes, the proton is a stable particle. Isolated protons seem never to change into anything else. But grand unified theories predict that the proton must decay eventually.

Most of the matter in the universe is hydrogen. The proton

is the nucleus of the hydrogen atom. Two or more protons are in the nucleus of every other atom. About half your body weight is protons. If the proton is unstable, then all matter is unstable. Nothing is permanent.

Different versions of the grand unified theory have different ways of estimating the proton lifetime. Most place it somewhere around 10^{38} or 10^{39} seconds. (For the rest of this chapter, time will be expressed in seconds, since the time line is in seconds and the spans would require exponential notation even if expressed in years.)

This is a half-life. It does not mean that every proton disintegrates 10^{38} or 10^{39} seconds after its formation. It means only that half of the universe's protons will have disintegrated in 10^{38} or 10^{39} seconds. *Some* protons should be disintegrating today, even though no proton (or anything else) is 10^{38} seconds old.

The proton has a positive electric charge. Charge cannot vanish even if the proton does. Theory predicts that the proton should decay in one of two ways. In each case, a positron is produced. The positron is the antimatter form of the electron. It has a positive charge and fulfills the requirement that the proton's charge not vanish. In one decay mode, the proton should produce energetic photons along with the positron. In the other mode, neutrinos would be produced.

If the proton really does decay, it should be possible to detect the positron and the photons. (Actually, the positron and a nearby electron would immediately be annihilated, producing photons. Photons would be detected in any case.) It isn't possible to wait 10^{38} seconds, but it is possible to monitor 10^{30} protons for a few years, which amounts to the same thing as far as proton decay is concerned. Experiments of this sort are going on now. A decay has been reported, but it may have been spurious.

In Dyson's original paper on the cosmological future, he explicitly excluded the possibility of proton decay. Other cosmologists, notably Duane A. Dicus, John R. Letaw, Doris C. Teplitz, and Vigdor L. Teplitz, have studied the consequences of proton decay in detail.

Proton decay would affect dense matter differently from interstellar or intergalactic masses. Let's take a rocky planet such as Pluto, which, we'll assume, has escaped the sun's fiery death and compression in a black hole and by 10^{38} seconds is still around, coasting through intergalactic space. Our time machine is traveling very quickly now, so that the far more than glacially slow process of proton decay is starting to happen all at once.

A proton in Pluto's interior turns into a positron and some neutrinos. The neutrinos pass through the planet without effect. The positron immediately encounters an electron—there is an electron for every proton in ordinary matter—and is annihilated, forming energetic photons. The photons are absorbed and reemitted by nearby atoms. This raises the temperature locally.

The decay also leaves the original atom with one less proton. It may well tear the nucleus apart; at the very least, it creates an atom of a different element, which may not be stable.

The net effect of many proton decays in a massive body is to maintain the body at a certain (low) temperature and to continually decrease its mass. Ultimately, a planet, rock, or dust grain would "evaporate" away to nothing.

Much the same thing would happen to dead stars. The equilibrium temperature due to proton decay for a dead star is estimated to be only a few degrees above absolute zero. For a neutron star, the temperature might be about 100 degrees Kelvin.

These temperatures are very cold by today's standards, but proton decay might be the principal source of heat in the distant future. All fusion will have ceased. All the conventionally radioactive substances will have decayed long ago. The microwave background radiation is currently at an intensity that suffices to heat up objects—even in intergalactic space—to about 3 degrees Kelvin. But the temperature of the background radiation is decreasing as the universe expands and will be negligible by 10^{38} seconds. If protons do decay, stars will still be hotter than their surroundings.

Intergalactic gas is so thin that the average atom almost never encounters another atom. Most of the atoms are hydrogen, a proton electrically bound to an electron. It might be thought that the positron emitted in proton decay would immediately annihilate the electron, with the result that hydrogen atoms would simply vanish.

Calculations of Dicus, Letaw, D. Teplitz, and V. Teplitz imply that this will not happen. Both electrons and positrons are infinitesimal compared to the distance between them. The positron is ejected out of the atom at high speed, and the chance of it encountering its own electron is about 1 in 100 million.

For dense matter, this means that a positron must pass through about 100 million neighboring atoms, on the average, in order to strike an electron and annihilate. For atoms in intergalactic space, there are no neighboring atoms. A thin gas or plasma of positrons and electrons results. The plasma is made of half matter and half antimatter and yet it does not annihilate. The distances between the particles are too great; the effective sizes of the particles are too small.

After proton decay has run its course, the universe will consist of electrons, positrons, neutrinos, photons—and black holes. The black holes may then account for most of the mass. There will be no atoms, no solids, no liquids, no objects perceptible by any human sense. There will be no chemistry and certainly no chemistry-based life. All the complexity of our world will evaporate away. Grand unified theories predict that even atoms are a passing phase. Assuming the theories are right, the age of atoms will run from 10^{12} to 10^{38} or 10^{39} seconds—about 26 or 27 orders of magnitude.

Whether protons decay or not, there are events yet more remote in future time. Black holes are not entirely stable. Quantum theory demands that they slowly radiate away their mass-energy.

All this is theoretical. No one has ever seen a black hole, so the prediction that they radiate slowly is all the more removed from observation. Nonetheless, black holes are demanded by general relativity, and black-hole radiation is demanded by the uncertainty principle. These theories have proved correct in all cases where it has been possible to test them. Physicists are more confident that black holes exist and slowly radiate than they are about proton decay, for instance.

The matter that falls into a black hole is crushed to infinite density, losing all its identity. A black hole has a certain total mass that increases every time something new is added to it. If you drop a one-pound weight into a black hole, the black hole's mass increases by one pound (less any mass that is shed as radiation during the fall). This increase is manifested in the gravitational field of the black hole. In principle, you could put a satellite in orbit around the black hole, measure its orbit before and after adding the weight, and determine that the gravitating mass of the black hole had increased by one pound.

It was originally thought that there was no way that any sort of mass-energy could exit a black hole. A black hole was literally black because nothing, not even photons of visible light, could escape from it.

Stephen Hawking showed that if this were entirely true, black holes would violate the uncertainty principle. Physicists are not prepared to dismiss the uncertainty principle, even when dealing with black holes. Hawking showed that the conflict could be resolved if black holes slowly radiate their mass away as photons.

The process is so slow that it makes no perceptible difference under most circumstances. A black hole such as might form in a supernova explosion or in the center of a galaxy is still a perfectly black void into which matter may fall but not escape. Every now and then such a black hole would emit a photon, but it would take billions of years for it to lose measurable mass. Meanwhile, most black holes would sweep up enough matter to more than make up the difference.

Radiation by black holes can only become important in the long, long run. Even by 10^{38} seconds, it will not have been a significant process. The process speeds up as the mass of the black hole decreases. A black hole with the mass of the sun takes about 10^{71} seconds to evaporate to nothing. But much of the mass-energy is radiated in the very last second of the black hole's existence.

By then the black hole is very different from the usual picture of a black hole. The diameter of a black hole is directly proportional to its mass (not to the cube root of mass, as might be expected). As the mass of the black hole accelerates toward zero, the black hole shrinks ever faster, ultimately becoming far smaller than a proton. As it shrinks, the black hole emits radiation at an increasing rate. In its last second, a black hole emits 10^{31} ergs of radiation. Far from being black, it is an intense point of light that appears suddenly and vanishes even more suddenly.

It is barely possible that mini-black holes were formed in the big bang. They might have the mass of a mountain or a planet, and some might be ready to radiate out of existence even now. Barring this possibility, it is unlikely that black holes of much less than a solar mass could form by any process operating in the universe today. The lightest black holes will be the first to evaporate, so black holes should start vanishing when the universe is about 10^{71} seconds old.

Black holes of larger masses will vanish over many orders of magnitude on the cosmic time-scale. Dying black holes are sometimes called "white holes" for their presumed dazzling brilliance (which will have no competition as all the stars will be long dead). The largest black holes, those in the centers of galaxies, are estimated to take 10^{107} seconds to become white holes and vanish.

Individual white holes are fleeting, but as a class they will be one of the most permanent fixtures in the universe. The age of white holes runs from 10^{71} to 10^{107} seconds, or 36 orders of magnitude. In comparison, stars are found only over 6 orders of magnitude, atoms over 26 or 27 (assuming proton decay), and life seems confined to about 5.

That takes the cosmic time-scale up to 10^{107} seconds or 10^{100} years—to the year googol. By then the universe will be a thin gas of photons, electrons, positrons, and neutrinos, the electrons and positrons too far apart to annihilate. Meanwhile, the universe will have been expanding all this time. By one estimate, the average distance between particles will be greater than the diameter of the observable universe today. If this picture of the far future is accurate, then all physical processes will have run their course by 10^{107} seconds. The universe will have stabilized.

·10·

RANDOM FIELDS

Random fields are the Life universe's big bang. Life's recursive rules ferment a primordial chaos to produce structure. The evolution of structure from a random field is simpler and better understood than real-world cosmology. Life random fields caricature the generation of complexity in our world.

The only distinguishing feature of a random field is its density. The density can be interpreted as the probability that any given pixel is on at time 0. The fate of a random field is easy to predict only for the two extreme cases. If the density is 0, every pixel will be off and stay off forever. If the density is 1, every pixel will be on and switch off a moment later from overpopulation.

If the density is near either extreme, the results are almost sure to be the same for a small finite field, such as the screen of a home computer. Any on cell has eight neighbors. If the density is much less than one-quarter, the chances of it having the two on neighbors necessary for survival are slim. Similarly, a sprinkling of off pixels in an otherwise all-on field will probably not prevent total die-out.

Intermediate densities are the most interesting on a home computer. Generation 1, the first to have felt the rules of Life, is often sharply denser or less dense than the random initial state. A low initial density drops further in generation 1, as many cells die from isolation. A density of about one-third increases to about 40 percent in generation 1. An initial density of about 40 percent does not change much in generation 1. Higher densities drop from overpopulation.

164

For intermediate densities, the first few generations look like low-resolution video snow. The pattern changes, but it still looks random. Few or no recognizable Life objects are seen. The density slowly decreases. Gaps appear and grow. As time goes on, stable objects appear. Some active regions are isolated as density drops.

COMMON UNSTABLE OBJECTS

Certain active regions occur again and again in random fields. They therefore qualify as natural "objects." In every case, these common active regions can arise from small predecessors.

Several unstable objects have already been mentioned—the T tetromino, the R pentomino, the B heptomino, the pi heptomino. All turn out to be common in random fields. Think of an unstable object as an abstract movie. Each generation is one frame of the action. Names such as T tetromino refer to one frame, really, but are also used to describe the entire movie.

Many other patterns converge on the T tetromino's evolution. All are loosely called T tetrominos, even though the true T tetromino pattern never appears in some histories. In isolation, every such evolutionary path must terminate in a traffic light. In a random field, interaction with other objects may prevent this (as happens in the R pentomino's history). The terminology is further blurred so the term *traffic light* is also used to describe this evolution, even when the constellation of four blinkers does not occur.

A welter of patterns also converge on the histories of the R pentomino, B heptomino, pi heptomino, and Herschel. All these patterns grow so large and take so long to stabilize that their mature constellations are virtually never seen in a random field. Something always interacts. But these unstable objects can be recognized from their early generations. The pi heptomino is easiest to spot because of its symmetry. Many of its early generations suggest hearts or arrowheads. The illustration shows an evolving random field with two pi heptominos (the smaller is generation 4 of the true pi heptomino). Both will soon interact with other, more chaotic active regions. The field

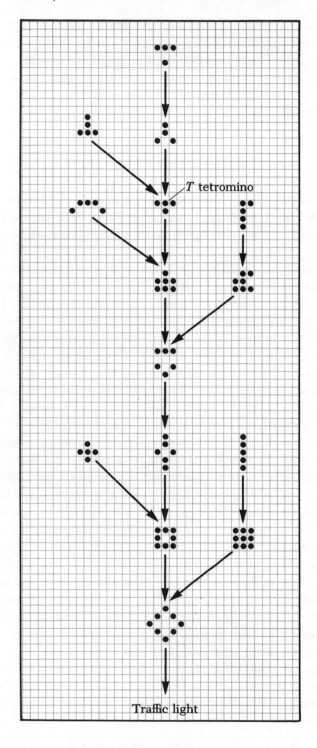

T tetromino

Traffic light

also contains a mature traffic light, blocks, beehives, blinkers, gliders, and loaves.

Not only the traffic light but the other symmetrical constellations are preceded by a characteristic evolution. So the term *honey farm* is broadened to include the floral pyrotechnics that precedes the constellation, as well as the various small predecessor patterns.

There are other, less common symmetrical constellations. "Lumps of muck" is a loose constellation of four blocks. It has only one degree (rotational) of symmetry. It forms from the stairstep hexomino in sixty-three generations. Generation 5, shown in the random-field illustration, is a pair of off-center, generation-2 pi heptominos. All the stairstep hexomino/lumps of muck's generations retain this symmetry.

A "fleet" is a compact constellation of four ships. The ships touch end to end in pairs. Strictly speaking, a pair of touching ships ought to qualify as a still life in its own right. It is sometimes called a "shiptie," in which case a fleet is a pair of shipties. The fleet arises in about a dozen generations from predecessors that include a ship touching an extra pixel diagonally.

"Interchange" is a stretched version of a traffic light with six blinkers rather than four. It forms from a glider collision or from two beehives touching end to end. Two traffic lights form, but two of their blinkers are so close that they annihilate. The "bakery" is a symmetrical constellation of four loaves. The loaves actually touch in pairs to form fourteen-pixel "biloaves."

Other common unstable objects never acquire any symmetry. "Century" is a common object that takes a hundred generations to stabilize. One small predecessor is an *R* pentomino with an extra pixel. Century's final constellation is a row of three blocks, not quite evenly spaced, and a blinker off to the side. A toad forms in century's evolution but is destroyed before maturity.

The "block and glider," named for its final constellation, arises from a six-pixel pattern resembling a foreshortened eater. It, too, takes about a hundred generations to stabilize.

The histories of these unstable objects are intertwined. In Chapter Two it was noted that the honey farm, traffic light, and Herschel evolutions all occur in the *R* pentomino's evolu-

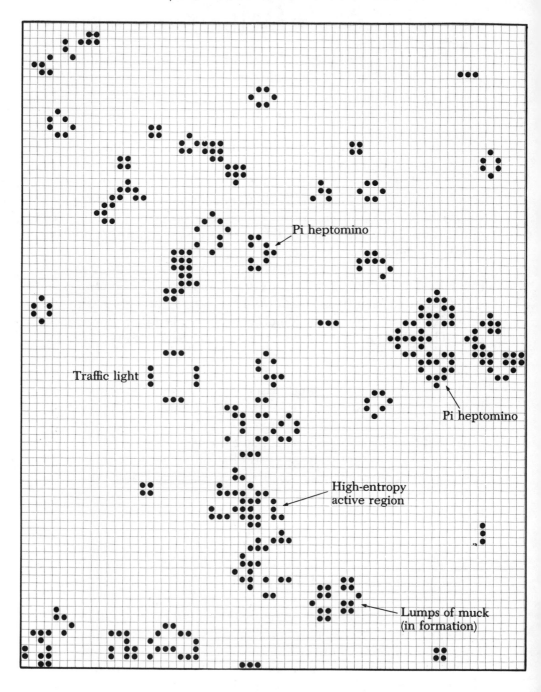

Pi heptomino

Pi heptomino

Traffic light

High-entropy
active region

Lumps of muck
(in formation)

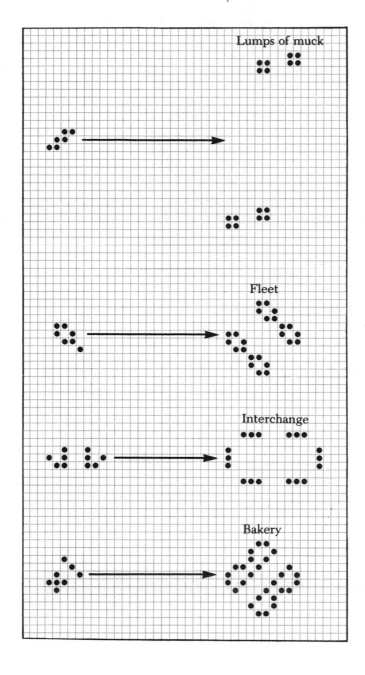

Lumps of muck

Fleet

Interchange

Bakery

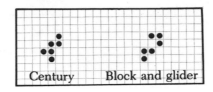

Century   Block and glider

tion around generation 48. The *B* heptomino occurs earlier in
the *R* pentomino's evolution. The normal evolution of the *B*
heptomino produces a block and Herschel; the *B* heptomino
is in fact the source of the block and Herschel that appear in
the *R* pentomino's generation 48. Lumps of muck and the
shuttle (without the stabilizer blocks, it is an unstable object)
also occur in the *R* pentomino's long evolution. Indeed, the
R pentomino occurs in its own evolution.

The *R* pentomino's evolution avoids being infinitely regressive. Later generations include a replay of some early generations, but the new *R* pentomino does not have unlimited space
in which to evolve. It eventually interacts with the debris and
ceases to evolve as an *R* pentomino. This is the same thing that
happens to the pi heptomino in its evolution. These objects are
like movies in which some of the action takes place in a motion
picture theater so that other movies are visible within the
movie. Some of these other movies, in turn, may contain clips
from still other movies. Occasionally, a movie may contain
clips from itself. But no movie can contain all of itself within
itself (provided there is some regular action in addition to the
movies within movies).

Other histories are equally incestuous. The bare shuttle
reacts with the beehives it produces and splits into two centuries, which evolve to maturity. The *T* tetromino/traffic light
evolution figures in the formation of the interchange, pulsar,
and pentadecathlon and in the pi heptomino's evolution. The
early generations of the acorn include *B* and pi heptominos;
later generations include the fleet and the usual traffic lights
and honey farms.

All these unstable objects are special in that they come from
small predecessors. The symmetry of some betrays that fact. A
pattern is likely to acquire symmetry only as a small object.
Once it becomes too large, dissimilar parts of the pattern are
almost certain to evolve differently.

Experienced Life enthusiasts grow familiar with the evolution of the common unstable objects. An evolving fleet or pi heptomino is no less predictable than a block or blinker. However, there is more to the random field than stable and common unstable objects. There are also high-entropy active regions.

A high-entropy active region is one without a small predecessor. It takes a large amount of information to describe it precisely. The random-field illustration on page 168 has a large, high-entropy region at lower center. In general, any miscellaneous active "junk" has high entropy. The initial random state is *all* high-entropy.

High-entropy active regions can feed on stable objects. An active region that interacts with a stable object seems more likely to remain active than an isolated active region. In general, the active regions of a largely stabilized field take longer to die out than they would without the background of still lifes and oscillators.

FREQUENCIES OF LIFE OBJECTS

The block and blinker are the two most common objects in random fields. The frequencies of other common Life objects have been tabulated. The chart plots the frequency of occurrence of objects versus their size.

These data are based on work done by Hugh Thompson. Thompson tracked over eight thousand small patterns to stabilization. Thompson's frequencies seem to agree well with the frequencies of objects in random-field experiments. In general, the objects in random fields arise from small patterns that separate out. Of course, the larger constellations are unlikely to reach maturity intact in a dense random field.

The block is used as the standard. Its frequency of occurrence is set equal to 1. The frequency of everything else is a number less than 1. Only the few most common objects approach the block in frequency. In order to show less common objects, a logarithmic scale is used.

The beehive is about 45 percent as common as the block. The *lone* blinker is next (40 percent as common), and the

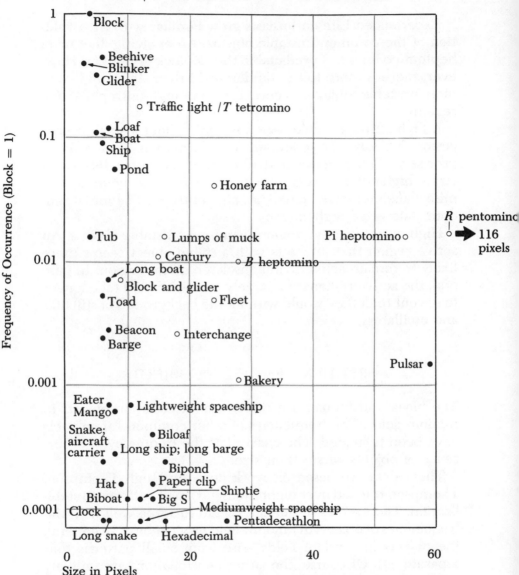

glider next (32 percent). In a random field, gliders eventually strike something, of course.

The traffic light constellation is 18 percent as common as the block in Thompson's data. If you prefer to count a traffic light as four blinkers, then this frequency should be multiplied by four and added to the frequency for lone blinkers. Then blinkers are somewhat more common than blocks. But many traffic lights do not reach maturity in random fields. It is more reasonable to count traffic lights separately and to use Thompson's figure to represent, approximately, the frequency of the *T* tetromino/traffic light evolutionary path.

The next most common object is the loaf (12 percent), followed by the boat (11 percent), the ship (9.1 percent), and the pond (5.6 percent).

There is no simple rule for determining which objects are common and which are not. In general, small objects are more common than large objects, but there are plenty of exceptions. The tub is as small as any still life. Yet it is only about 1.6 percent as common as the block. On the average, you see over sixty blocks for every tub in a random field.

The blinker is the only common oscillator. The toad and beacon are roughly a hundred times rarer. The six-pixel clock, near the bottom of the chart, is less than one ten-thousandth as common as the block. But the pulsar, with 48 to 72 pixels, is nearly as common as the beacon.

Likewise, the glider is far, far more common than any other spaceship. Lightweight spaceships are about as common as eaters or mangoes. The middleweight spaceship is about ten times rarer, and the heavyweight spaceship is too rare to show up on the chart.

Some of the more common unstable objects are charted. The *R* pentomino, pi heptomino, lumps of muck/stairstep hexomino, century, *B* heptomino, and block and glider, are all about 1 percent as common as the block. They are plotted by size of their final constellations, even though most do not reach maturity in a dense random field. Herschel is evidently more common in the *B* heptomino and *R* pentomino evolutions than on its own and is not plotted.

The chart shows some larger still lifes not mentioned yet. They are the "bipond," the "hat," the "paper clip," the "big S," the "biboat," and the "hexadecimal." Like the ship-tie and biloaf, the bipond and biboat are paired still lifes, touching but not interacting. The hexadecimal is disjointed, a beehive resting on a "table." Although much less common than the block, these still lifes are the commonest of their size.

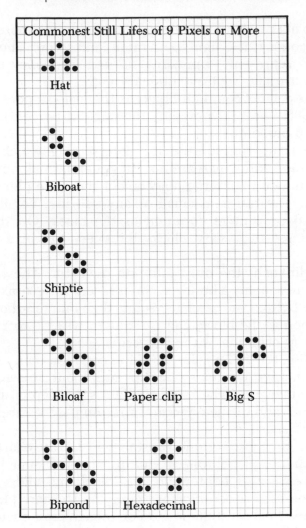

Commonest Still Lifes of 9 Pixels or More

Hat

Biboat

Shiptie

Biloaf   Paper clip   Big S

Bipond   Hexadecimal

DO ALL FIELDS STABILIZE?

All finite fields eventually stabilize. The active regions recede and vanish, leaving a constellation of stable objects only. It is easy to prove that this must be the case. If x is the number of pixels in the display, then 2^x is the number of possible configurations. 2^x is generally an astronomical number, but it is certainly finite. Unprecedented patterns cannot keep turning up endlessly.

The larger the field, the longer it will take to stabilize. There is no guarantee that an infinite field ever stabilizes. Search large enough areas of the field, and you will always be able to find active regions.

The average final density of a random field is about 5 percent in typical random-field experiments, which are usually done with intermediate starting densities. If the field starts with a density of much less than 30 percent or much more than 40 percent, the final density is usually less.

Speculation about "living" Life patterns focuses on infinite, low-density random fields. Consider the fate of a field that starts with a vanishingly low density—say, one in a billion pixels on. This experiment can be done only in imagination. No display contains a billion pixels, so any screen would probably be blank even in generation 0. What's more, virtually all on pixels of generation 0 have no neighbors and die in generation 1.

Not quite everything vanishes. There must be rare instances of two adjacent pixels, even at one-billionth density. There must even be ultrarare instances of three pixels together.

Many arrangements of three pixels die out. The only way they can survive is for at least two pixels to be in an orthogonal row, with the third pixel in any of the six positions orthogonal to the two-pixel unit. Four of the six positions create pre-blocks; the other two create blinkers.

After generation 1, the field must be composed mainly of blocks and blinkers in a 2 to 1 ratio. All the other groupings of three or fewer pixels will have vanished.

This is not to say that objects other than blocks and blinkers do not form in a very thin field. They do, but they are far rarer. The beehive, traffic light, pond, and tub require four pixels to form. If the density is very low, the chance of four nearby pixels must be negligibly small compared to the chance of three. Similarly, the objects that require five pixels must be exceedingly rare compared to the four-pixel objects. The relative frequencies in the chart must not apply to very low density fields. In such fields, the frequencies of objects must be stratified by the size of minimum predecessor patterns. Any possible Life objects can occur, but large objects would be astronomically rare.

The lower the density, the greater the average spacing between objects. In a thin enough field, gliders could travel miles before striking anything. R pentominos could evolve into their constellations without interference. If there are self-reproducing Life patterns, they would have room to grow in such a field.

·11·

VON NEUMANN AND
SELF-REPRODUCING MACHINES

John Von Neumann is a fascinating, paradoxical figure in twentieth-century thought: original, prolific, multitalented—and villainous, some would say. He has never been well known to the public, but several of his ideas have shaped the modern world. As much as any other American, perhaps, Von Neumann was responsible for the development of the electronic digital computer. His "theory of games" has influenced economic and military thought, spawning such buzzwords as "zero-sum game." At Los Alamos, Von Neumann advocated the "implosion method," the ultimately successful design for the fission bomb. He rallied support for an accelerated hydrogen bomb program in 1950. At that, most of Von Neumann's work—and certainly much of the most innovative—was in pure mathematics.

Von Neumann was born in Budapest, Hungary, on December 3, 1903. His father was a banker. The family was wealthy enough to have a country home and for the father to purchase a noble title, as was the custom at the time. Von Neumann was recognized as a prodigy almost from the time he could speak. As a child, he could divide two eight-digit numbers in his head. He entertained family guests by memorizing columns from phone books, then reciting names, addresses, and phone numbers unerringly. One of the anecdotes about the young Von Neumann has his mother pausing during crocheting and staring into space a few moments. "What are you calculating?" he asked.

177

At the age of twenty, Von Neumann published a formal definition of ordinal numbers that has been used by mathematicians ever since. In his mid-twenties, Von Neumann noticed a startling connection between quantum physics and vector theory. He discovered that the states of quantum systems could be represented by vectors in an abstract, infinite-dimensional space. By the time he immigrated to the United States (1931), he was one of the most important living mathematicians.

The excursion into quantum physics was a watershed. Von Neumann's study of vectors and quantum states had advanced both mathematics and physics. Increasingly, Von Neumann's attention turned to applied mathematics.

In 1931 Von Neumann became professor of mathematics at Princeton University; in 1933 he was appointed to the nascent Institute for Advanced Study. Von Neumann and his two wives are remembered as Princeton's leading socialites. For years they gave the town's largest parties, often with touches such as an ice sculpture of Princeton's experimental computer. Von Neumann had something of a reputation for aphorisms— "Only a man from Budapest can enter a revolving door behind you and emerge ahead of you," for example. He liked rich foods and drink and expensive furnishings. While at Los Alamos, he would drive 120 miles to a favorite Mexican restaurant. He is credited with having been able to drink a quart of rye whisky in an hour and drive. He wrecked cars almost annually, yet never sustaining serious injury. Above all, Von Neumann was famous for working at odd times—in taxis, during nightclub floor shows, while waiting for breakfast. He would slip away to his study during parties, leaving the door open because he liked the sound of parties.

Through the 1930s and early 1940s, Von Neumann worked on game theory. Game theory is an idealized study of games among rational opponents. Von Neumann had hoped it would be the nucleus of a future exact science of economics. Game-theoretic analysis requires that all motives be prespecified. In that sense, game theory is a highly artificial model of human conflict. But Von Neumann realized that for many situations this does not matter. In ticktacktoe, checkers, chess, and poker, the only object is to win. As long as players agree to play to win, motives can be abstracted away.

After playing ticktacktoe a few times, most people realize that the game is pointless. If both players act wisely, the game must end in a draw. A winner can emerge only when one or both players makes a strategic mistake. Von Neumann was able to prove that games such as chess are "pointless" in the very same way.

Von Neumann did not specify the "correct" strategy for chess. That would require giving the countermove for every possible move at every stage of the game. The quickly branching tree of moves and countermoves soon grows unmanageable. Instead, Von Neumann and colleague Oskar Morgenstern showed that all two-person games where one person wins and the other loses ("zero-sum" games, with no cooperation possible) and where no relevant information is hidden from either player ("perfect information" games, as opposed to card games where players cannot see the other players' hands) must have an unbeatable strategy for one or both players.

An unbeatable strategy does not guarantee a win. Both players may have unbeatable strategies. Then any rational game must end in a draw, as in ticktacktoe. Of course, when only one player has an unbeatable strategy, that player can always win. The only asymmetry between players in chess is that the white player is allowed to move first (or to pass the first move to black). Consequently, white can be at no disadvantage. Von Neumann's game theory predicts one of the following. Either white always wins in fully rational chess, or a rational game always ends in a draw. No one knows for certain which is the case.

Game theory became a vogue among economists and military strategists. It has remained an object of controversy. The abstract content of game theory is not in dispute. The question is whether basic theorems of game theory can have any practical relevance to far more complex "games" such as the steel industry or NATO defense. Critics have claimed that the mindset of game theory reflects Von Neumann's axiomatic cynicism as much as real mass behavior. "It is just as foolish to complain that people are selfish and treacherous as it is to complain that the magnetic field does not increase unless the electric field has a curl," runs another of Von Neumann's epigrams. "Both are laws of nature."

From about 1940 on, Von Neumann was in demand as a

consultant to industry and government. He became wealthy. In May 1954, for instance, he was a consultant to twenty-one organizations, including the air force, the CIA, IBM, Los Alamos Laboratories, the Pentagon, and Standard Oil. In his autobiography, Ulam writes that Von Neumann "seemed to admire generals and admirals and got along well with them. . . . I think he had a hidden admiration for people or organizations that could be tough and ruthless." Some colleagues felt that Von Neumann's increasingly worldly interests were at the expense of his mathematics. But his seminal work with automata theory—some of his most celebrated mathematics—was an outgrowth of his military consulting.

Von Neumann supervised the design of unprecedentedly powerful computers for the American military during World War II and after. While wrestling with practical problems, Von Neumann became interested in the potential abilities of automatons. He was particularly impressed with the work of British mathematician Alan Turing. Turing showed that a relatively simple computer can perform any possible calculation, given the right programming. Von Neumann wondered if a relatively simple robot could perform any physical task, including the construction of duplicates of itself.

Von Neumann pondered the problem of self-reproduction from time to time but found himself busy with defense work. After the war, most of the scientists and mathematicians who had worked at Los Alamos returned to academia. Von Neumann remained strongly political, his views reflecting an ingrained fear of the Soviets. "This was definitely a three-way war," he said at the Oppenheimer hearings. "I considered Russia an enemy from the beginning . . . and the alliance with Russia as a fortunate accident that two enemies had quarreled."

Biographers have attributed Von Neumann's attitudes toward the Soviets to his background. As a Hungarian and a Jew, he had historic reason to distrust the Russians. But by 1950, his politics were verging on paranoia.

Von Neumann came to support the now-almost-forgotten doctrine of "preventive war." On August 29, 1949, the Soviet Union exploded its first fission bomb in Siberia. The American nuclear monopoly was over, but plainly the Soviets had few if

any functional bombs on hand immediately after the proto-
type explosion. The advocates of preventive war held that
America should seize the moment to bomb the Russians back
to a new "stone age."

In 1950, Von Neumann was pushing for an immediate first
strike on all major Russian cities and military targets. He argued with the conviction of a chess strategist that the Soviet
Union must be annihilated before it developed intercontinen-
tal missiles and nuclear stockpiles. *Life* magazine quoted Von
Neumann, not disapprovingly: "If you say why not bomb them
tomorrow, I say why not today? If you say today at 5 o'clock,
I say why not one o'clock?"

Few Americans supported preventive war. Truman ignored
the matter. However, at the urging of Von Neumann, Edward
Teller, and Admiral Lewis Strauss, Truman initiated an accelerated hydrogen bomb program.

Von Neumann's last years were spent on a variety of projects. He worked on the hydrogen bomb. He became an unoffi-
cial government spokesman in behalf of atomic testing, de-
nouncing Linus Pauling's warnings about cancer risk from
radioactive fallout. He worked on his proof that a machine can
reproduce, searched for patterns in the digits of pi, and busied
himself with more arcane mathematics. Von Neumann played
with a visionary scheme to dye the polar icecaps and change
world climate. The dye would absorb the sun's radiation and
melt the ice. Iceland would become as balmy as Hawaii, and
somehow computer modeling would forestall disaster. In 1955
Eisenhower appointed Von Neumann to the Atomic Energy
Commission. By then Von Neumann was no longer advocating
preventive war—the Soviet nuclear arsenal was too big.

The same year, Von Neumann fell on a slippery floor and
hurt his shoulder. The doctor examining him found a malig-
nant bone cancer, already spreading to other parts of the body.
Nine years earlier, Von Neumann had witnessed the Operation Crossroad atomic tests on Bikini atoll. In retrospect, it
seems likely that his cancer resulted from the radiation expo-
sure at the Bikini tests. Von Neumann continued to work as
Atomic Energy Commissioner for many months after the diag-
nosis of cancer. A lifelong agnostic Jew, he converted to Ca-
tholicism in his hospital bed. He died in Washington, D.C., on

February 8, 1957, attended by security-cleared air force or-
derlies in case he mentioned military secrets in his delirium.

THEORY OF SELF-REPRODUCING AUTOMATA

Von Neumann's analysis of machine procreation is typical of
the way he blended pure and applied mathematics. Is his proof
an exercise in pure logic? Mathematics applied to biology?
Mechanical engineering?

Von Neumann did not finish his proof. His manuscript trails
off into a series of notes about the remainder of the proof,
written at various times when other projects did not interfere.
A colleague, Arthur W. Burks, completed the proof in Von
Neumann's spirit and had it published as *Theory of Self-Reproducing Automata* (Urbana and London: University of Illinois
Press, 1966).

Von Neumann first formulated a *kinematic* model of reproduction. He imagined a robot floating in a lake. Also floating in the lake are all the components needed to build the
robot. He pictured the robot collecting needed components
as it jostled against them and assembling them into a copy of
itself.

Among the components in the lake were "kinematic" elements—artificial arms that could be used to move other objects. There were "fusing" elements to join two components
together and "cutting" elements to separate them. "Sensing"
elements located needed components. "Girders" provided a
rigid framework for construction. Also, girders could be joined
together like Tinkertoys to build an external memory for the
robot's computer brain.

The external memory was important. The robot's brain
would have to be rather sophisticated. It would have to be able
to distinguish components, judge their distance, decide when
to grab them, what to do if a component slips out of a kinematic element's grasp, how to correct for the rocking motion
of objects when the water is choppy, and so on. Although the
robot was in no way expected to exhibit intelligence, the design of its computer brain would still pose formidable problems. Von Neumann did not want to tackle these problems in

detail. Instead, he fell back on Alan Turing's notion of a "universal computer."

Turing had shown the remarkable fact that a fairly simple computer can perform any possible calculation at all. With the right programming, a "universal Turing machine" can do everything the most powerful computer ever built, or ever to be built, can do. If the human brain is itself an immensely complex computer, as seems to be the case, then a Turing machine can do anything the human brain can do.

There is nothing magic about a universal Turing machine. Although usually intended as a mathematical abstraction, a Turing machine could be designed and built from the usual electrical components. The only catch is that a Turing machine must have an infinite memory capacity—or what amounts to the same thing, it must be able to add on to its memory as needed. This infinitely extensible memory will be outside the main body of the computer, so it is called an external memory.

Turing pictured his universal computer hovering over an infinite strip of paper tape marked off into squares. The squares contained 0s or 1s. Any possible program for the computer could be encoded as a string of 0s and 1s on its tape. Likewise, the computer could erase and print its own 0s and 1s to keep track of intermediate results of calculation—use the tape as scratch paper. In long computations, the computer might well use light-years of tape. At the end of a calculation, the answer would be expressed as a coded string of 0s and 1s.

Turing's discovery is not as unbelievable as it may seem. It is something like the observation that it is possible to build a fairly simple machine (the typewriter) that can produce any possible novel. Everyone realizes that a novel lies mostly in knowing what keys to push and in what order, and hardly at all in the typewriter.

So it is with the universal Turing machine. The Turing machine must first be given a complete set of instructions on how to perform the desired calculation. The price of its utter versatility is that the machine is little more than a blank slate. Everything must be spelled out for it.

It is easy to design a typewriter for the English language. All possible words and sentences can be constructed from a few letters, numerals, and punctuation marks. It is difficult to design a Mandarin Chinese typewriter. There is a different character for each word. In practice, Chinese typewriters include only the most common words. There are Chinese sentences that cannot be typed on a Chinese typewriter.

Turing's discovery was that logic has a structure rather like the typography of a phonetic language. There is a small number of basic logical manipulations. Any calculation whatsoever can be constructed from them. A machine can perform any of these basic operations—or all of them. Given the right instructions, a universal computer can compute anything.

If the desired calculation is very complicated, then the instructions will probably have to be very long. The universal computer may take a long time to come up with an answer. It may never find an answer, if the problem is unsolvable and beyond the power of any computer. But the universal computer (aside from its external memory) need be no different or more complicated to handle a difficult task than a simple one.

There are many ways of designing a universal computer. The details are not important so much as the basic simplicity. Von Neumann's machine needed to be able to figure out how to manipulate its environment so as to make copies of itself. Sometimes the problems it encountered might be easy; sometimes they might be difficult. The question was whether a single fairly simple machine could handle these problems in all cases.

Turing's work showed that it could—provided the self-reproducing automaton contained a universal computer. For the requisite external memory, Von Neumann used girders. The robot constructed a half-ladder of girders with some of the rungs missing. A missing rung stood for 0; a rung was a 1.

Von Neumann essentially succeeded in showing how his floating machine could reproduce. Yet he was dissatisfied. He had hoped to capture the logical essence of self-reproduction. Unfortunately, much of his analysis was bogged down with the problems of motion in his nonexistent lake.

It was infuriatingly difficult to move components to where they were needed. True, living organisms have comparable problems—a protein molecule must be digested and its component amino acids distributed to the cells of the body. But Von Neumann's lake was far enough removed from biochemistry as to have no real interest of its own. Further, these problems of motion, which occupied so much of Von Neumann's attention, were basically unrelated to the real issues of self-reproduction.

The lake and the floating gizmos were a fiction anyway. Von Neumann took the abstraction one step further.

THE CELLULAR MODEL

Von Neumann adopted Ulam's suggestion. He started thinking in terms of a cellular array evolving by recursive rules. Each cell of an imaginary checkerboard receives input from its neighbors and then decides what to do in the next instant of time.

The recursive rules are the physics of the cellular space. Von Neumann chose his rules so that information could be stored and manipulated much as in our world. After all, it was the information theory of self-reproduction that interested him. The physics shouldn't be too wild. Many simple patterns should stay put and not change. Evolution of simple patterns should not contain any surprises. In these respects, Von Neumann's cellular space is very different from Life.

In particular Von Neumann wanted to avoid those problems that had plagued his kinematic model: locating a needed component out of a random soup, reeling it in, and moving it to a construction site. Things would be so much simpler if the robot could wish components into existence at the very spot they were needed. Von Neumann decided to allow just that in his cellular model. He designed his recursive rules so that new "matter" could be created in response to pulses sent out by the robot brain.

I will not go into the details of Von Neumann's proof. Conway has shown how to translate the essence of it into the simpler Life array. It will suffice to show Von Neumann's general line of reasoning.

Each cell in Von Neumann's space can have any of twenty-nine states. One state can be thought of as the off or empty state much as in Life. That leaves twenty-eight distinct ways for a cell to be on.

Von Neumann wanted to show that there are patterns of cells that reproduce themselves in an otherwise empty array. The reproduction would be nontrivial. The patterns would contain a complete coded description of their own organization, a blueprint. They would contain a universal constructor. They would also contain a universal computer to coordinate the process.

Basically, reproduction would go like this. A pattern would read its own blueprint. It would send the information to its universal constructor. The universal constructor would grow a "constructor arm"—a peninsula of active cells—out into an empty region of the plane. The "head" of the constructor arm would scan back and forth, producing new on cells in its wake. A new copy of the self-reproducing pattern would be built scan line by scan line, much like the image on a television screen.

Again, Von Neumann did not attempt an explicit proof. That is, he did not find a specific pattern that met the requirements of self-reproduction. Finding such a pattern would be a tedious, mostly unenlightening task. Von Neumann was busy enough with design minutiae for real computers.

Instead, Von Neumann sought to show that every necessary component of a real computer or robot constructor could be duplicated by a pattern in his cellular space. He further had to show that every design problem could be solved. Most important, he had to show how a finite configuration could encapsulate a complete description of itself.

Every possible three-dimensional computer or electrical contraption can be (and generally is) represented by a two-dimensional circuit diagram. The diagram consists of a finite number of components and a finite number of connections between components. The states of Von Neumann's space were chosen to facilitate the conversion of a circuit or logic diagram to an equivalent cellular pattern.

Generally speaking, some wires must cross on a circuit diagram. This is shown with a symbol:-⌃-. The symbol means that the wires do not touch in three dimensions but must be shown crossing in a two-dimensional diagram. Among the design problems Von Neumann had to and did solve was to show that the "wires" of his cellular computer could cross without confounding their signals.

Von Neumann had to show that every basic component of a universal computer was possible in his cellular space. A universal computer requires only these elements:

- AND gates: Two input wires lead to an AND gate. If they both contain a pulse at a given moment, then the AND gate sends a pulse through its output wire. If only one or no pulse arrives through the input wires, there is no output pulse. Be it a mechanical relay, a vacuum tube, a solid state device, or a pattern of cells in an abstract space, an AND gate embodies logical conjunction. An output is produced only if input 1 *and* input 2 contain pulses.
- OR gates: Two input wires lead to an OR gate. If one *or* the other—or both—contains pulses, the OR gate produces an output pulse.
- NOT gates: The NOT gate produces an output pulse only if there is *not* a pulse on its input wire. It fails to produce an output pulse only if there is a pulse of input.

By wiring up enough components of these three types, it is possible to build a universal computer. Von Neumann designed AND, OR, and NOT gates out of his cells. He showed how to connect gates with arbitrary time delays should it be desired. He showed how to add on to an external memory, which was a sort of Turing-machine tape made of two types of cells. He designed a universal constructor.

THE CENTRAL PROBLEM OF SELF-REPRODUCTION

Of all the design problems Von Neumann tackled, one stands out as the most important and most universal. It is the problem of avoiding infinite regress in the blueprint. Living organisms are finite and reproduce in a finite time. Any machine, whether in the real world, Von Neumann's cellular space, or Life's cellular space, is likewise finite. If self-reproduction involves infinities, then it is pointless to look for self-reproducing machines.

The problem with machine reproduction is that the universal constructor is a mindless robot. It has to be told very explicitly what to do. The more things that have to be spelled out, the longer the description or blueprint must be. The longer the description, the less chance that everything that has to be said can be expressed in a finite blueprint.

Feed a description of an alleged self-reproducing machine to its universal constructor. Literal-minded as it is, the universal constructor does not understand that it is supposed to be reproducing the machine, *including* its description. Instead, it turns out a new machine without description. The temptation is to kick the machine, curse its lack of initiative, and finish the job yourself. All you need do is to copy the description and give the copy to the new machine.

Wait! No special creativity is required to copy a blueprint and hand the copy over to the new machine. Every step of the process can be precisely spelled out, which is to say, a machine could be built to do it. Von Neumann's solution was to append a "supervisory unit" to the machine to handle these latter tasks. The supervisory unit makes the machine more complex, of course. The blueprint must be enlarged in order to show the supervisory unit as well as the universal constructor.

Consider what happens when a blueprint of the universal constructor-plus-supervisory unit is given to the universal constructor-plus-supervisory unit. Under the guidance of the supervisory unit, the universal constructor builds a new universal constructor-plus-supervisory unit. When the construction is complete, some sort of internal switch clicks on in the supervisory unit. The supervisory unit goes into a new mode of operation: It copies the blueprint and transfers the copy to the new machine. In short, a machine (universal constructor-plus-supervisory unit) plus a blueprint of itself gives rise to a new machine plus a blueprint of itself.

Infinities are avoided because the blueprint does not try to encapsulate itself. The blueprint describes the universal constructor and supervisory unit. When it comes time to make a new blueprint, the blueprint *is* its own blueprint. The blueprint does double duty; it is interpreted successively in two ways. It is first interpreted literally—as a set of directions to be followed in order to make a certain type of machine. Once the supervisory unit switches into its second mode, the instructions in the blueprint are ignored. The blueprint becomes merely a raw material for the copying process.

Von Neumann's solution is a neat trick, but it is much more general than that. It is the method by which real, organic life reproduces.

By the 1940s, the basic chemicals of life had been identified: proteins, fats, nucleic acids, and their simpler constituents. The large-scale structure of nucleic acids in cells (the double helix) was not known. Nor did biochemists have much of an idea how these particular chemicals fitted together and interacted so as to reproduce themselves.

Starting with Watson and Crick's discovery of DNA's structure, discoveries in biochemistry have followed the general framework of Von Neumann's cellular machine. There is, first of all, a blueprint in the form of DNA. A cell's DNA contains a complete description of all essential parts of the cell *except* for the DNA itself (no infinitely regressive blueprints).

There is a universal constructor in the form of ribosomes. The information in DNA is relayed to the ribosomes via messenger RNA. The ribosomes use this information to build proteins.

The notion of a universal constructor is a little confusing; it is a machine that can build "just about anything" given the right instructions. In the case of the ribosomes, the "just about anything" means any protein molecule that can be constructed by linking the twenty or so amino acids that can be coded in DNA and RNA.

Protein molecules are strings of amino acids. The string is often many thousands of amino acid units long. In principle, it can be as long as desired. So the possibilities are limitless. Silk, collagen, albumen, and keratin are common examples of the types of molecules that the ribosomes can construct.

There is generally a limit to the versatility of a universal constructor. No possible DNA instruction could induce ribosomes to produce polyester, fiberglass, or steel. How, then, do we decide whether a constructor is universal?

The most important criterion is the ability of the constructor to participate in nontrivial self-reproduction. The ribosomes qualify, but in a somewhat indirect way.

Ribosomes can make only protein. But there are important

components of the cell that are not protein. DNA is not a protein. Nor are RNA, ATP (an important cellular energy source), the heme in human blood hemoglobin, cellulose in wood, or phosphates in bone.

The trick the ribosomes use is to make protein intermediaries to handle construction they cannot perform themselves. Special kinds of proteins—enzymes—construct needed nonprotein molecules in the cell. In a broader sense, the ribosomes plus the enzymes throughout the cell are the cell's universal constructor.

It is also a set of enzymes that forms the bulk of the cell's supervisory unit. An enzyme called DNA polymerase acts as the cell's blueprint copier. Other enzymes determine when copying will take place and switch the ribosomes on and off. The only basic difference between organic and Von Neumann's reproduction is that Von Neumann arbitrarily pictured his machine as duplicating its blueprint after the machine itself had been duplicated. In living cells, duplication of DNA takes place early in the replication process. The construction of new cellular "machinery" is an ongoing process, continuing even after the cell has split in two.

The exact correspondence to biology was not known at the time of Von Neumann's work, but his proof had a strong philosophical impact on biologists. Von Neumann showed that there was no magic in self-reproduction, that the exact process could be spelled out and programmed into a machine with a certain minimum level of complexity. By itself, Von Neumann's work made no specific assertions about biology. Reproduction of living organisms might still involve an imponderable life force. Von Neumann did, however, strike down the argument that self-reproduction *must* proceed by supernatural means.

AN INFORMATION-THEORY DEFINITION OF LIFE

Assuming that biological reproduction is ultimately mechanistic, Von Neumann's analysis can be used to formulate a definition of life in terms of information. In information theory, the physical structures encoding information (the printed page, electric pulses in wires, videotape, etc.) mean nothing; only the information itself counts. The aspects of life that seem important in information theory may be different from those that seem important in other fields. To a biochemist, complex carbon compounds are the salient feature of life. To a behavioral biologist—and to most people—the ability to react to stimuli is crucial. To an ecologist, the ability to increase exponentially in the absence of predators and abundance of food is characteristic.

All these attributes are mere details from the information-theory standpoint. The truly remarkable features of life are just those elucidated by Von Neumann:

(1) A living system encapsulates a complete description of itself.

(2) It avoids the paradox seemingly inherent in (1) by not trying to include a description *of* the description *in* the description.

(3) Instead, the description serves a dual role. It is a *coded* description of the rest of the system. At the same time, it is a sort of working model (which need not be decoded) of itself.

(4) Part of the system, a supervisory unit, "knows" about the dual role of the description and makes sure that the description is interpreted both ways during reproduction.

(5) Another part of the system, a universal constructor, can build any of a large class of objects—including the living system itself—provided that it is given the proper directions.

(6) Reproduction occurs when the supervisory unit instructs the universal constructor to build a new copy of the system, including a description.

These features can even be adopted as a definition of life. All the usual forms of life meet these criteria. But the cogency of this information-theory definition of life is better seen when applied to problematic cases. Whenever biologists try to formulate definitions of life, they are troubled by the following: a virus; a growing crystal; Penrose's tiles; a mule; a dead body of something that *was* indisputably alive; an extraterrestrial creature whose biochemistry is not based on carbon; an intelligent computer or robot.

No reasonable person would maintain that a crystal or Penrose's tiles are alive. Nor would a reasonable person fail to count a mule as alive, or even an extraterrestrial creature of exotic biochemistry that is in all other ways lifelike. The other examples are more open to interpretation. Certainly no formal definition of life should go against the most familiar notions of what is alive.

A virus consists of a DNA or RNA core surrounded by a protein coat. The DNA or RNA encodes the amino acid sequence needed to construct new copies of the protein coat. But once inside a cell, the DNA or RNA can serve as templates for the construction of new genetic material. Thus a virus is a system that contains a complete description of itself. It meets criteria (1), (2), and (3).

Beyond that, it falls short. A virus does not contain the finely tuned system of enzymes that functions as a supervisory unit. Nor does it contain a universal constructor. Viruses manage to reproduce only by commandeering these systems from the cells they attack. By the information-theory definition, an isolated virus is not alive. It does not have the self-reproductive sophistication of Von Neumann's machine.

Crystals and Penrose's tiles are similar phenomena. Neither a crystal nor a two-tile unit contains a description of itself. Their simple growth does not involve the manipulation of information; they are not alive.

A mule is sterile yet plainly alive. The essential self-reproductive machinery of all earthly life is in the cell. Think of a mule as a colony of cells. Each cell contains a complete DNA description of the undifferentiated mule cell, as well as a supervisory unit and a universal constructor. Even if the mule does not reproduce its kind, the mule cells do. They meet the criteria of nontrivial self-reproduction.

A similar argument applies to a dead body. An animal that has just died still contains some living cells with all the self-reproductive machinery. Conceivably, a bioengineer might be able to clone a living animal from a cell taken from the dead body. So it is reasonable to maintain that living matter becomes nonliving matter only as the information and construction machinery are lost.

It might seem like stretching a point to confer living status on mules or dead bodies, based on reproductive potential. Reproduction seems to be tangential to these systems. But systems as complex as a mule could never have arisen on earth if it had not been for nontrivially reproducing precursors. Biochemists believe that the first pre-cells on earth arose from nonliving matter and that these pre-cells had only the minimum of organization necessary for reproduction. Ecological pressures favored better-adapted pre-cells. Nontrivial self-reproduction allowed acquired modifications to be preserved in DNA for future generations. After billions of years' worth of these cumulative modifications, one of the results was the mule. The mule, like all other familiar forms of life, is almost entirely the product of long-iterated self-reproduction.

Information theory would also deem a Martian of alien biochemistry to be alive. Again, the relevant point is the message and not the medium: If genetic information and universal constructors can be embedded in some other chemistry, then the resulting systems would be just as alive.

Trickiest of all the cases is a sophisticated computer or robot. The presumed potential of computer engineering confounds any attempt at a behavioral definition of life.

Part of the problem with a behavioral definition is that many forms of life have no behavior to speak of. Some bacteria and spores do not exhibit irritability, assimilate nutrients, or eliminate wastes for long periods. And bacteria are not anomalies: They are representative of the oldest, most common forms of life.

Furthermore, it seems likely that it is possible to build a computer or robot to perform any precisely definable task—including whatever tasks one may claim are the earmarks of life.

Could a computer be alive? Using a reproductive definition, the answer is yes—but not on the strength of reaction to outside stimuli or "intelligence." Von Neumann's early kinematic model of machine reproduction was a literal, nuts-and-bolts robot. His cellular model was no less a robot, though in an abstract universe. Both reproduce nontrivially and are alive by the information-theory definition.

A computer could be arbitrarily powerful and *not* be alive, however. If it is just a computer and does not have the means for reproduction (a coded description of itself, a robot universal constructor, etc.) then it fails to meet the criteria. A supersophisticated computer might be intelligent, or even "conscious" in some sense, without being alive by this definition.

Perhaps, then, a distinction should be made between "reproductive-alive" and "sentient." Humans, and the animals most familiar to us, are both. Bacteria and all plants are only reproductive-alive. Maybe computers could be sentient only.

Still, humans are descended from distant ancestors that were reproductive-alive only. Computers are built by humans. In practice, sentience always can be traced to nontrivial self-reproduction—even in those cases where it seems divorced from reproduction.

THE COMPLEXITY BARRIER

Von Neumann sometimes spoke of a "complexity barrier." This was the imaginary border separating simple systems from complex systems. A simple system can give rise to systems of less complexity only. In contrast, a sufficiently complex system can create systems more complex than itself. The offspring systems can beget more complex systems yet. In principle, any set of physical laws that permits complex systems allows an unlimited explosion of complexity.

(Exact) self-reproduction is a feature of systems right on the complexity barrier—systems that preserve but do not increase their level of complexity in their offspring. Von Neumann was interested in seeing exactly what minimum level of organization is necessary for self-reproduction. As one index of that minimum complexity, he estimated the size of minimal self-reproducing patterns in his cellular space. Burks, J. W. Thatcher, and others have refined some estimates.

It is possible to design a universal constructor that just fits in a 57-by-143-cell rectangle. That comes to 8151 cells, of which many are in the off state. Given the correct sequence of pulses, this 57-by-143-cell pattern sprouts a constructor arm, sweeps out a pattern of new cells, and then withdraws the constructor arm back into the pattern.

The universal constructor turns out to be a small part of the self-reproducing pattern—about 4 percent, by one estimate.

The supervisory unit and parts concerned with manipulation of memory are larger. Much of the pattern is given over to units that code and decode information transmitted to or received from other parts of the pattern. Von Neumann estimated the total size of his self-reproducing pattern at 200,000 cells. The shape can vary depending on design.

Von Neumann's proof treats some related issues as well. Can a machine make a machine more complicated than itself? It can, Von Neumann concluded. The universal computer embedded in the self-reproducing machine can perform any finite calculation whatsoever. It follows that it can be programmed to read the blueprint and then design a machine that is more complicated in any precisely specifiable way. Each generation of machines would be more complex than any of its predecessors. This is approximately what has happened (very slowly) in biological evolution. No life force is required for evolution of more complex organisms, either.

Von Neumann's cellular space is contrived: He designed the twenty-nine states and recursive rules with the express purpose of making it easy to construct patterns that reproduce nontrivially. Since then, other mathematicians have devised simpler cellular spaces and shown that patterns can reproduce in them. The ultimate reduction is to just two states. (Nothing can happen in a cellular array with just one state for cells.) That Life's two states permit nontrivial reproduction is even more remarkable than that Von Neumann's space does. Life's rules were not designed to facilitate reproduction. The existence of self-reproducing Life patterns is strong evidence that the type of self-reproduction Von Neumann imagined is a natural phenomenon, possible in many contexts.

·12·

SELF-REPRODUCING
LIFE PATTERNS

The name *Life* has come to have a double significance. Conway's name referred only to the way the growth or death of small patterns suggests the growth or death of biological populations. Gosper's discovery of the glider gun convinced Conway that there is a lot more to Life. Over a period of years, Conway translated Von Neumann's cellular self-reproducing machine into the simpler Life universe. Life players talk of "living," self-reproducing Life patterns.

What would a self-reproducing Life pattern look like? Would it be alive? Conway published his version of Von Neumann's proof in *Winning Ways (for Your Mathematical Plays)* (New York: Academic Press, 1982), an immense work covering many games other than Life, coauthored with Elwyn Berlekamp and Richard Guy. Conway concluded: "There are Life patterns which behave like self-replicating animals. . . . It's probable, given a large enough Life space, initially in a random state, that after a long time, intelligent self-replicating animals will emerge and populate some parts of the space."

Conway's proof is detailed. It is founded on the realization that the glider gun and many other Life objects can be constructed in glider collisions. Conway demonstrated that vast constellations of glider guns and eaters can produce and manipulate gliders to collide in just the right way to form a copy of the original constellation. Conway's proof incorporates Von Neumann's reasoning about self-reproducing machines and

196

machines that make machines more complex than themselves. Because of its importance, this chapter describes every step of Conway's argument. You may want to skim this chapter first and then come back to it. No step of the argument is difficult, but a number of different ideas are used.

LIFE COMPUTERS

The first part of Conway's argument deals not with reproduction *per se* but with a Life computer. Conway showed that there are Life patterns that function as universal computers.

Let's clarify this point. Life is generally played on a computer, a real-world computer of wires and microchips. Conway, however, is talking about Life patterns—images of on and off pixels on a sufficiently big video screen—that act like computers. In principle, you could start with a computer pattern, and a pattern representing its programming, and have it calculate any desired result.

The result would have to be expressed as a Life pattern, too. If it was a numerical computation whose answer turned out to be 7, perhaps the Life computer could fire off seven gliders or arrange seven blocks in a prespecified display area.

Conway had the romantic notion of Life patterns that model open questions of mathematics or logic. He used the example of Fermat's last theorem, a famous conjecture of number theory. Fermat claimed to have discovered a proof of the (would-be) theorem but never wrote it down. No mathematician has since proven the theorem true (or false, for that matter). Logicians have demonstrated that there is no general method for deciding the validity of a mathematical statement. It is often possible to find a proof that a statement is true or false, but there is no magic formula that is guaranteed to decide the issue. The truth of Fermat's last theorem remains undecided.

Conway showed that it is possible to design a Life computer to search for a proof that Fermat's last theorem is false—and vanish entirely if and when it accomplishes that task. Similar Life computers could work on any other undecided issues of mathematics. It follows that the question of whether a given Life pattern ever vanishes may be no less undecidable than the unresolved issues of mathematics and logic. (If not, building a Life computer would be a magic formula for testing the truth of any and all mathematical statements.) Life is inherently unpredictable.

The basis of Conway's Life computers is the glider stream. The fundamental construction is a glider gun firing a stream of gliders into an eater. Such streams may be positioned at right angles so that they intersect. Conway showed that it is possible to weave a universal computer out of thousands of intersecting glider streams. The only additional component needed is the block, used as an external memory register.

Real-world computers encode information as electric pulses in wires. The patterns of pulses symbolize binary numbers— a pulse encodes a 1; absence of a pulse encodes a 0. Everything a real digital computer does, it does with binary numbers. For the present discussion, it is not even necessary to ponder why this is so. Let it be conceded, on the basis of the success of contemporary computer design, that any necessary information can be encoded and manipulated in binary-number form.

A glider stream can encode any binary number. Each glider of an evenly spaced stream represents a bit. Remove gliders to encode 0s; leave them in to represent 1s. The resulting coded stream is the exact equivalent of the pattern of electric pulses in a regular computer's wires.

The distance between gliders in the stream that emerges from a glider gun is 7½ pixels. That poses a severe problem. If the streams are the wires of a Life computer, they must be able to cross without interfering. That isn't possible for two period-30 streams. Two new-gun (period-46) streams can be positioned to interpenetrate. That is scarcely better, though. A given stream may have to cross hundreds of other streams— some without interaction and some with interaction. The only way to guarantee that any desired circuit can be implemented is to use guns of arbitrarily long period. The longer the period, the more room between gliders and the more flexibility in orchestrating streams.

THE THIN GUN

One of the glider collisions is surprisingly useful. It has its own name: the "kickback reaction." The kickback reaction is a 90-

degree glider collision in which the only product is a glider. The product glider travels in a direction opposite to one of the original gliders. It is as if this glider has been reflected or kicked back. Besides being reflected, the product glider is shifted by half a diagonal space. The process takes six generations.

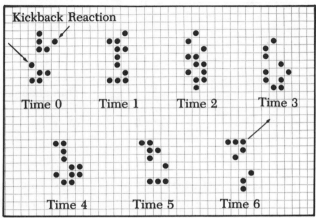

Creating a long-period gun also requires a collision in which both gliders vanish. Mutual annihilation is the most common outcome of glider collisions, so there are many such collisions to choose from.

It is possible to build a gun of arbitrarily long period from three ordinary glider guns, an extra glider, and an eater. This type of gun is called a "thin gun" because its glider stream is thinned out to any desired degree.

Gliders travel at a 45-degree angle to the horizontal. Western graphics, book design, and thinking favor right angles. For clarity's sake, most of the diagrams in this chapter will be tilted 45 degrees so that glider streams are horizontal or vertical. Rather than show a pixel-by-pixel diagram of a thin gun, a simple schematic diagram of the type used by Conway will suffice. The three guns are symbolized by circled *G*s. The eater is a circled *E*. Glider streams are shown by arrows or by a row of little boomerang shapes. This simple diagram symbolizes an eater eating the gliders from a gun, for instance.

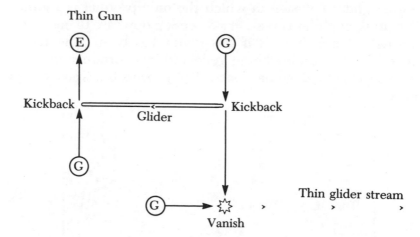

Thin Gun

In the thin gun, two ordinary, period-30 guns shoot parallel streams of gliders in opposite directions. A single glider shuttles between the two streams; its path is shown by the thin loop. Things are coordinated so that each time the shuttling glider encounters one of the glider streams, it collides with a stream glider in the kickback reaction.

Specifically, it is the shuttling glider that is kicked back. The stream glider is annihilated. In the kickback, the shuttling glider's path is shifted slightly as well. The kickback on the left shifts the path up; the opposite kickback on the right shifts the path down. Hence the closed loop.

Let's say that the spacing is such that the shuttling glider encounters the left glider stream every 3000 generations. Then it interacts with 1 out of every 100 period-30 gliders. The stream glider that kicks back the shuttling glider is destroyed. So the left glider stream above the kickback site has every hundredth glider missing. It could be symbolized as a string of 99 1s and a 0.

The same is true of the right glider stream below the kickback site. The left glider stream is not needed and is fed to an eater. But the right glider stream meets still a third glider stream at right angles in a vanish reaction.

The bottom gun and the right gun are timed and spaced so that their gliders annihilate each other. Every 3000 generations, the right stream has a glider missing. The glider from the bottom gun that was supposed to annihilate with the missing glider passes through unharmed. The result, at lower right, is a thin glider stream that has only a hundredth as many gliders as normal.

The thinning factor (100 here), it turns out, must always be divisible by 4. You can build a thin gun with ¼ as many gliders as usual, ⅛ as many, ¹⁄₁₂, ¹⁄₁₆, etc. By placing the two parallel glider streams far apart, the thinning factor can be as high as desired.

The thin gun solves the problem of wire-crossing. If the basic, information-carrying glider streams are thin enough, then there will be enough space between bits for as many other streams as may be necessary to cross. So thin guns rather than simple guns will be the basic guns of the computer.

Every computer has a clock, a device that sends electric pulses to all parts of the computer to initiate the next basic operation. A thin gun will be the clock for our Life computer. The computer will be very slow. Conway guessed that a thinning factor of 1000 might suffice for a simple self-replicating pattern.

There is one more requirement for circuit design. Gliders travel only in straight lines. The wires of a circuit diagram must often turn corners. A Life computer requires some way of reflecting gliders 90 degrees. This was one of the most worrisome problems Conway faced. Its resolution uses some of the same design tricks useful for the rest of the computer.

AND, OR, AND NOT GATES

The three fundamental logic gates are easy to implement with glider streams and vanish reactions. The inputs to these logic gates are thinned, coded glider streams, already carrying information.

A Life AND gate has two input streams. They are timed so that bits (the positions in the thinned stream that may or may not contain a glider) arrive at the AND gate in perfect synchronization. The two input streams and the gate's output stream can be represented with 1s for gliders and 0s for vacant bits:

Input stream A	0001011101
Input stream B	1101000100
Output stream	0001000100

A glider appears in the output stream only if both input streams had gliders.

The OR gate has a glider of output when one input stream *or* the other has a glider:

Input stream A	0001011101
Input stream B	1101000100
Output stream	1101011101

The OR gate fails to produce a glider only when both input streams have empty bits.

Finally, the NOT gate negates its single input. It changes all the 0s to 1s and the 1s to 0s:

Input stream	0001011101
Output stream	1110100010

The NOT gate is simplest to construct; it was used in the thin gun. Have a coded glider stream meet an uncoded (111111111 . . .) stream at right angles in a vanish reaction. All of the gliders in the original, coded stream will annihilate. Gliders from the full-strength stream will escape intact only when a glider is missing from the coded stream. The resulting stream is the negation of the original.

For the AND gate, the two input streams come in parallel. A gun fires across both streams and into an eater. The gun's gliders meet the stream gliders in vanish reactions. The output stream is the continuation of the lower input stream.

How can a glider make it into the output stream? Only if it was in the original B stream—and only if it is not annihilated by a glider from the gun. It can avoid annihilation only by having a glider in stream A to remove a glider in the gun stream. In other words, only if A *and* B have glider pulses will the output have a glider pulse.

Stream A must be delayed relative to stream B by the distance between the two vanish reactions. This distance may be small or as great as desired. That allows a lot of flexibility, but still it requires that we be able to delay glider streams. The gun in the AND gate may be a simple, period-30 gun as its stream

NOT Gate

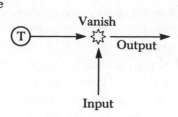

AND Gate

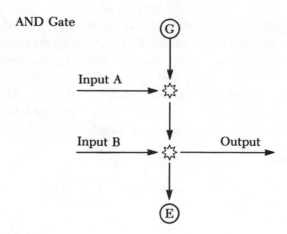

OR Gate

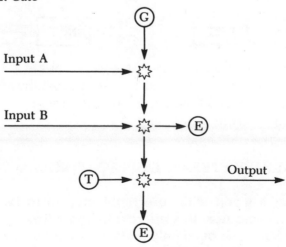

is ultimately consumed by the eater. The NOT gate requires a thin gun (circled T in the diagram) as its stream becomes the output after coding in the vanish reaction.

The OR gate is similar to the AND. A gun fires across two parallel streams of input. If either input stream removes a glider from the gun's stream, then a glider from a second, thin gun's stream escapes a vanish reaction and becomes part of the output stream. Two eaters mop up excess gliders.

Both the OR and the AND gates may require that one input stream be delayed. The NOT gate can be used to delay a stream. Just build a detour in the stream to be delayed, using NOT gates for the corners. There must be an even number of NOT gates so that the stream leaves the detour the same as it entered (but delayed). After the first NOT gate, the stream is negated (empty spaces for gliders and vice versa). After the second, it is returned to its original form; the third NOT negates it again; the fourth restores it again.

Delaying a Stream

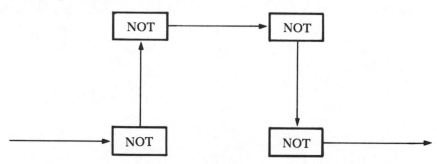

NOT gates can also turn a stream through a 180-degree angle or shift it into a parallel course. Unfortunately, there is no way that the NOT gate can turn a stream through a 90-degree angle without negating it.

A GLIDER STREAM DUPLICATING MACHINE

Sometimes a given wire must split and go to two or more places in a computer. In a microchip-and-wire computer, this is no problem. One need only make a Y-shaped junction in the wire (or its equivalent). But glider streams will not branch of their own accord. It is necessary to make a duplicate copy of the stream. A glider stream duplicating machine is required.

Several tortuous schemes for duplicating glider streams were devised. At length, Conway came up with a much simpler solution. Luckily, Conway's solution also allows streams to turn corners without being negated.

Conway's duplicating machine uses the kickback reaction to bounce gliders back headlong into their own stream. Take a glider stream that has been thinned the minimum of four times (a period-120 thin gun). A glider from a crossing stream kicks back the first glider in the period-120 stream. It smashes into the second glider in a 180-degree collision. The spacing is such that the result is a block. Then the third glider crashes into the block. In this case, both block and glider annihilate. The fourth and all subsequent gliders are unaffected.

The net result, then, is that a single glider can remove three gliders from a full transverse stream. The transverse stream must be thinned as a full period-30 stream does not allow enough space for the collisions.

The duplicating machine works as follows. A gun thinned four times fires across several streams into an eater. The computer's data streams will surely be thinned more than four times. For the sake of exposition, assume that the basic computer data is thinned forty times—so that the 4× thin gun fires ten times for every bit of data.

The diagram represents glider streams from the standpoint of the 4× thin gun. The full, uncoded stream from the gun is represented 111111111111 . . . A glider passes a given point every 120 generations.

The data streams are much, much thinner. The first nine period-120 spaces are always empty. Only the tenth space may (or may not) contain a glider. The thinned data stream that would normally be represented as 101 becomes 00000000010000000000000000001 as represented here.

The stream position that may contain a glider is represented by a question mark. Therefore, the coded input stream can be considered as a series of ten-position units of which only one position contains any information: 000000000?.

The input stream to be duplicated enters on the left. It is first passed through an OR gate. The other input to the OR gate is

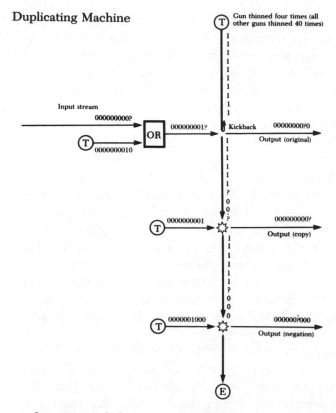

Duplicating Machine

a thin gun of the same period as the coded stream (thinned forty times here). The thin gun is delayed relative to the input, though, so that its glider occurs in a different position: 0000000010. The OR gate combines the two streams to produce a stream that *never* has a glider in eight positions, *always* has a glider in one position, and *sometimes* has a glider in the lead position (if the input stream did). This stream is represented 000000001?.

The information-carrying stream meets an uncoded stream from a period-120 thin gun (1111111111) in a kickback reaction—the rebounding glider backtracking on the 1111111111 stream. Now consider what happens:

If there is a glider in the ? position, it will kick back the first glider in the 1111111111 stream, creating a pile-up collision there. The information-carrying glider itself will be annihilated. However, the glider behind it will pass through. The output, then, will be 0000000010.

If there isn't a glider in the ? position, the first glider in the 1111111111 stream will pass by without incident. But there is always a glider in the next position of the information-carrying stream. It will kick back the next glider in the 1111111111 stream. Meanwhile, the exiting output stream will be empty (0000000000).

To summarize, if the stream on the left of the kickback site is 0000000011, the stream on the right will be 0000000010. If the left stream is 0000000010, the right stream will be 0000000000. The right stream's position encodes the same information as the original stream: 00000000?0. The only difference is that the information-carrying bit has been shifted one position.

Now let's follow the 1111111111 stream after the kickback. The kickback removes exactly three gliders. If there was a glider in the ? position, then it is the three leading gliders that are missing. If not, the leading glider is intact but the three behind it are gone.

The situation can be symbolized like this: 111111?00?. The six backstream gliders are definitely there; two gliders near the front are definitely missing. Notice that the leading glider will be present only if the leading position in the input stream is empty, and vice versa. So the leading position is the negation of the information-carrying input position. Negations are sometimes symbolized by a bar over a symbol, so the status of the leading position can be symbolized $\bar{?}$.

Similarly, the position three bits downstream will have a glider only if the input does. It is symbolized with a ?. The gun's stream has been imprinted with two copies of the input information, one negated.

Two thin guns and two vanish reactions strip off the information. One gun produces a stream with gliders in the leading position. Since it is a 40X thin gun, all the other positions are empty. The output stream on the right can have a glider only in the leading position, and then only if the $\bar{?}$ position in the vertical stream is empty. The leading position is the negation of the negation of the input's leading position—so it is the same: 000000000?.

The next thin gun is positioned to produce a glider in the fourth position from front. The vanish reaction negates the ?

position in the vertical stream for an output to the right of 000000?000. The vertical stream is then reduced to a meaningless 1111110000 and is fed to an eater.

Conway's duplicating machine has three outputs, two of them simple copies of the input and one a negative copy. The information-carrying bits are shifted slightly in two of the outputs. All three output streams are parallel to the input stream.

Think of the duplicating machine as an adapter plug. Outputs that are not needed need not be used. (Unused outputs should be fed to an eater.) If more copies of the input are needed, the outputs may themselves be used as the inputs of other duplicating machines.

The duplicating machine finally provides a way of turning corners. Feed an input stream into a duplicating machine. Then feed the negated, parallel output to a NOT gate. The gate negates the negation, while turning the stream through 90 degrees.

90-Degree Glider Reflector

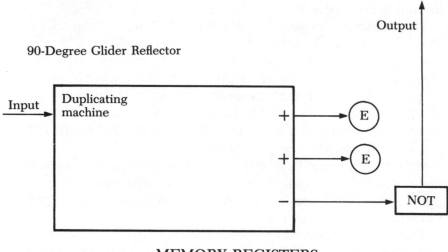

MEMORY REGISTERS

Wires and logic gates are all that any finite computer needs. A memory of any finite capacity can be built from these same elements. All it takes is a big circuit loop inside the computer. A long glider stream encoding any desired information travels this closed rectangular path. The four corners of the loop are the corner-turning devices. A duplicating machine is inserted in the loop so that the computer can skim off information without disturbing the memory. There are also OR and AND gates in the loop. Normally the computer sends a string of 0s to the spare input of the OR gate and a string of 1s to the free input of the AND gate. By proper manipulation of these gates, new information may be added to memory and old information erased.

A universal computer requires something more. It needs a potentially infinite memory. Von Neumann's cellular machine used a literal sort of Turing-machine "tape." The tape was a coded row of occupied and empty cells. There were, of course, an infinity of empty cells beyond the last occupied cell. In that way Von Neumann's machine realized a potentially infinite external memory. Von Neumann's machine grew an arm to read and modify the tape.

Computer scientist Marvin Minsky showed that a universal computer can use a somewhat different type of simple memory register. Minsky's registers must be able to store arbitrarily large numbers. The computer must be able to increase the number in a memory register by 1, decrease the number by 1, and tell if the number is 0.

Zero is the *only* register value that the computer can read immediately. Say a certain register holds the number 1,000,000. In order to read the number, the computer has to perform the zero test (which yields a "no" because the register value isn't 0), decrease the register by 1 (to 999,999), test for 0 again, decrease the register by 1 again, and so on. Finally, the register will be decreased to 0, the zero test will yield a "yes," and the computer will have the value 1,000,000 (encoded, perhaps, in an internal memory bank). The reading process always reduces the value of the register to 0. If the computer does not want to change the value in the register, it must immediately send 1,000,000 successive instructions to the register, each telling it to increase its value by 1.

Minsky's point is not that this is a particularly good way to build a computer. Rather, he proved that even this primitive way of organizing a memory suffices for all possible computations. Minsky's memory registers turn out to be easier to implement in Life than a Turing-esque tape.

Conway's idea was to use a block for a Life external memory register. The block would be outside the computer proper. Its

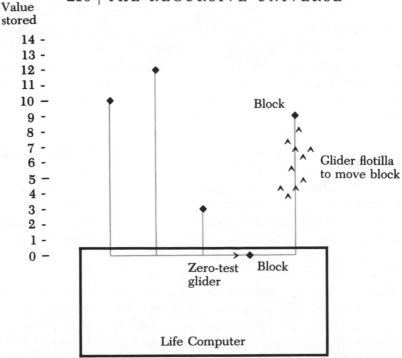

distance from the computer would represent the number being stored. Numbers of any size could be stored, since the block could be as far from the computer as necessary.

All this supposes that there is some way of moving the block even though it is not in the computer. Conway noted the fantastic variety of collisions in Life, and the fact that some collisions result only in one of the parent objects appearing to shift a little. Conway wondered if there was some way a glider could collide with a block, annihilating the glider but "pushing" the block farther out from the computer. Then there would have to be a complementary collision to "pull" the block back in.

All desired manipulations of the block should use gliders coming from one direction only—from the computer. Perhaps it would take more than one glider to do the trick, but they must travel in a parallel flock. If gliders are to be used to move the block, the block must move in a diagonal direction (shown as vertical in the diagram).

Conway examined the possible glider-block collisions. Because the block is symmetrical and a still life, there are only six distinct glider-block collisions.

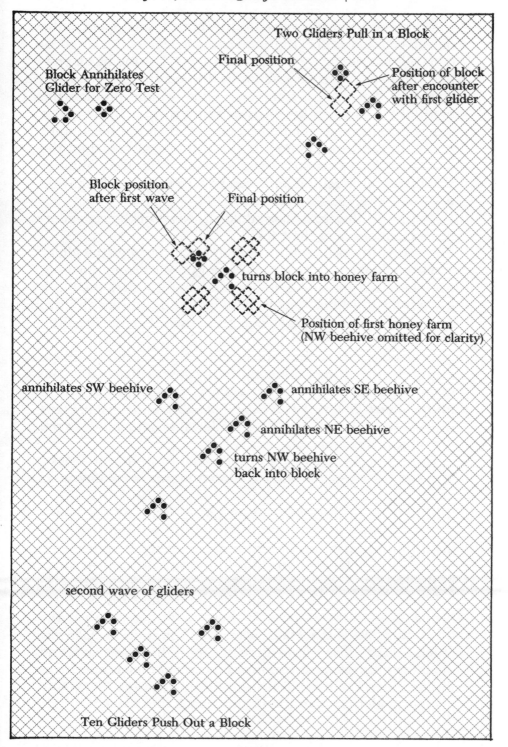

Two Gliders Pull in a Block

Final position

Block Annihilates
Glider for Zero Test

Position of block
after encounter
with first glider

Block position
after first wave

Final position

turns block into honey farm

Position of first honey farm
(NW beehive omitted for clarity)

annihilates SW beehive

annihilates SE beehive

annihilates NE beehive

turns NW beehive
back into block

second wave of gliders

Ten Gliders Push Out a Block

A glider can hit a block and destroy both (this happens in the pile-up collision used in the duplicating machine). An annihilation collision can be used for the zero test. Let the block position signifying zero be just inside the computer. Then the computer can shoot a transverse glider at the position the block occupies when the register value is zero. If the glider passes through and completes a circuit, then the block must not be in the zero position. If the glider vanishes, then the block must have been in the zero position. After the computer assimilates this information, it can restore the block by having two gliders collide the right way.

Moving the blocks is more difficult. None of the six glider-block collisions does the trick.

Conway was looking for a collision that destroys the glider and leaves only a block—in a different position from the original. There is a collision that does that, but it moves the block in a wrong direction. The block jumps like a chess knight, two squares in one direction and one in another. It comes in a little and shifts to the side.

A chess knight can move diagonally. A pair of mirror-image knight moves advance a knight three squares diagonally. The same thing can be done with a block. By shooting a pair of mirror-image gliders at a block, it is possible to move it in three pixels toward the computer. The sideways components of the two shifts cancel out.

Truly heroic measures are necessary to push out blocks. None of the six glider-block collisions seems to do any good at all.

One of the collisions creates a honey farm. As the honey farm blossoms out, parts get farther from the source of the glider. Unfortunately, the center of the honey farm is pulled in a little from the position of the block, and this cancels much of the benefit.

When the honey farm stabilizes, the farthest pixels of two beehives are just one pixel farther out diagonally than the block was. We don't really want the honey farm. Conway found a way to shoot three more gliders at the honey farm and annihilate the two near beehives and one of the far ones. This leaves a lone beehive a little farther out than the block. Still another glider changes the beehive back to a block.

We're not finished yet. After all the commotion, the block has again moved like a chess knight. This time, however, it has moved more to the side than in the direction we want. It takes another wave of five gliders to cancel out the sideways shift. The net result, after *ten* carefully choreographed gliders, is that the block is restored, just *one* pixel farther out diagonally.

Two gliders pull a block in three spaces; ten gliders push it out one. If a block position is to symbolize a number, the block must be pushed or pulled by the same distance. The simplest way to do this is to make three diagonal spaces the unit of distance. Then two gliders pull a block in a unit. Thirty gliders —three successive flotillas of ten—are needed to push a block out a unit.

This completes Conway's proof that Life patterns can function as universal computers—except for one detail that will be filled in in a moment. The circuitry of any possible computer can be mocked up in a Life pattern consisting only of guns, gliders, eaters, and blocks.

A LIFE UNIVERSAL CONSTRUCTOR

Life computers can produce any desired stream of gliders. Perhaps by crashing gliders together in just the right way, new Life computers can be built. The second part of Conway's proof shows how to construct any Life pattern that can result from a glider collision.

Chapter Four showed a two-glider collision that creates a block. Chapter Six showed a thirteen-glider collision that creates the gun. One of the two-glider collisions produces an eater (shown below). That leaves only the glider itself, and we can get all the gliders needed from the guns. Therefore, Life computers composed of Conway's four components can be formed in massive glider collisions.

Certain care must be taken. Eaters should be constructed before the guns. That way, once the guns start firing, the eaters will be ready to consume unwanted gliders. A collision need not be a near-simultaneous crunching of millions of gliders. It can be drawn out by putting space between incoming gliders. Construction of a Life computer could start with just two glid-

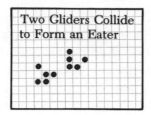

Two Gliders Collide to Form an Eater

ers meeting at right angles to form an eater. Then two more gliders on different paths form another eater. Gradually all the eaters would be built up, and then blocks for the guns and memory registers, and then finally the guns' shuttles. The incoming glider barrages could stretch far, far out from the collision site.

The collisions that form the gun, block, and eater all happen to be 90-degree collisions. That does not mean that a Life computer can be constructed from gliders coming from two directions only, however. Certainly the circuits of the computer will have to run in all four glider directions. That requires that guns and eaters be aligned in each of the four directions. Orientations of guns and eaters are determined by the collisions that form them. Consequently, gliders will have to come from all four directions.

How does a single Life computer arrange to have gliders converge from all directions? No matter what the design of the parent computer, it must fit completely within a rectangular region of the Life plane. Once its gliders leave that rectangular box, the computer can have no further influence on them. Left to themselves, gliders can only recede ever farther into the void. No collisions of a single pattern's gliders seem possible.

Conway and many others pondered this problem for several years. For a time, it looked as if Life might allow universal computers but not universal constructors. In that case, no Life object could reproduce. Finally, a wonderfully ingenious solution was found in the form of "side-tracking."

This discussion has glossed over a serious design problem of Life computers. Look at the glider flotillas used to move blocks in memory registers. They contain gliders moving down parallel paths, and the paths are only a few pixels apart. Now think about it: Every glider has to have been created in a gun somewhere (even if it's the gun in a NOT gate used to reposition a glider path). Glider guns are fair-sized objects and are aligned approximately diagonally to the streams they produce. It follows that of two gliders on close parallel paths, one must have been created first, and it—seemingly—had to pass through the gun that created the second glider. But chaos would erupt if one gun really did try to fire through another.

Conway found it is possible to use the kickback reaction to reposition gliders. The kickback is the two-glider collision that reverses one glider *and* shifts its path by half a diagonal space. (The glider's zig-zag path makes it possible to speak of a half-a-diagonal-space shift.) The thin gun does not make use of the path shift. The looping glider is kicked back by glider streams moving in opposite directions so that the path shifts keep canceling out.

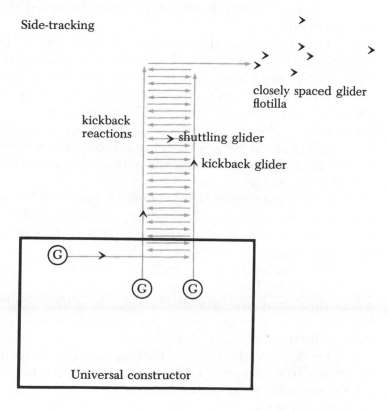

A shuttling glider can also be kicked back successively by glider streams traveling in the same direction. After the second kickback, the shuttling glider has been displaced by a full diagonal space; after the third, by a diagonal space and a half; and so on.

Conway called this technique side-tracking. Think of it as using three computer-controlled guns, two firing in parallel and one at right angles. All three guns are inside the universal constructor. The gun at right angles to the other two shoots a glider through an empty place in the nearer stream. Then the glider is kicked back by a glider in the farther stream. It doubles back, and is now kicked back by a glider in the first. For a while, both parallel guns produce periodic streams so that the glider weaves back and forth, moving a hair-width up at each kickback. Eventually, the shuttling glider is far away from the constructor, yet it is still imprisoned between the two glider streams.

At a time preordained by the constructor, one of the two parallel guns fails to fire. The shuttling glider escapes and continues on its path. Meanwhile, the first gun may well have sent other gliders up the same convoluted path. When these gliders escape, they will be traveling parallel to the first, along paths that may be as close as a few diagonal spaces apart. Getting the three guns to fire at just the right times requires some arithmetic, but since Life allows universal computers, any arithmetic problem is solvable.

More than one triplet of side-tracking guns will be needed to allow all possible spacings of gliders. For greatest flexibility, a Life computer-constructor should have a battery of triplets of side-tracking guns for each of the four glider directions. Then closely spaced parallel fleets of gliders can be produced anywhere in the Life plane outside the computer. Of course, the farther away the site of the desired glider flotilla, the longer it will take to get gliders out there—all the zig-zagging makes side-tracking very slow.

It takes four parallel glider flotillas, coming in from four directions, to construct a Life computer. Side-tracking allows a Life computer to assemble these glider flotillas—but that's not quite enough.

There are still restrictions on the directions of the glider flotillas. Even with side-tracking, the gliders must, basically, be moving away from the constructor. A 90-degree collision of two parallel glider flotillas is possible, but not a four-way collision from all directions.

Conway's remedy—and the final link in his proof—is "double side-tracking." Let three guns send a glider up far away from the computer and release it in a path parallel to the "edge" of the computer. Then have this glider kick back another glider fired out from the constructor from still a fourth gun. The result is a glider traveling *toward* the computer.

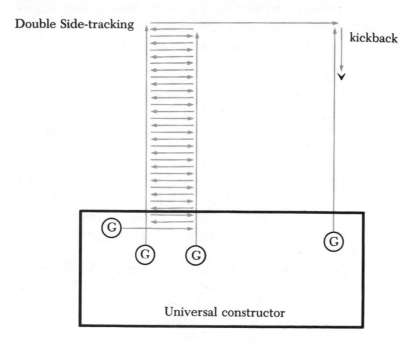

Double Side-tracking

kickback

Universal constructor

You wouldn't usually want a glider shooting into the computer, of course, but perhaps it could collide with other gliders to form some desired construction before smashing into the computer. It's easy to see that double side-tracking allows the computer to assemble any glider flotilla traveling in any direction at all. As much as ever, timing is crucial, but that is mere arithmetic. By careful planning, the computer can further ensure that there will be no stray gliders escaping to infinity from the side-tracking streams.

A SELF-REPRODUCING LIFE PATTERN

Given enough time, a Life computer can produce the four glider flotillas necessary to replicate itself and have them zero in on any desired location in the (otherwise empty) Life plane. Conway had to invoke many gimmicks to prove that this is so; let's see how they fit together.

First of all, what would a self-reproducing Life pattern look like? It would be big. Certainly it would be bigger than any video screen or computer in existence could encompass. It would also be mostly empty space. Design considerations dictate the use of very, very sparse glider streams. If you examined a portion of the interior of the computer closely, you would find only an occasional glider and, rarely, a gun or an eater.

There is no way of saying what the grand overall shape of the self-reproducing pattern would be. At large scale, it could assume any shape, if only you chose to design it that way. The pattern would have at least one type of external projection. A special set of blocks would reside at various distances outside the pattern's computer. From time to time, the computer would fire small flotillas of two or thirty gliders at the blocks to change their distance from the computer.

The blocks are the pattern's external memory registers. At least one of the blocks is special: It is the blueprint of the self-reproducing pattern.

More exactly, it is the *number* represented by the *distance* of this block from the pattern that is the blueprint. How can a single number represent the blueprint for a complex structure? One need only describe unambiguously every part of the structure in any convenient language (English, Spanish, FORTRAN, the machine language of the pattern's internal computer). The symbols of the language can all be encoded as numbers, and the numbers can be joined together into one big number. The block is placed this number of units from the zero position of its register.

Of course, if the blueprint is encoded in English, then the pattern's internal computer would have to be so sophisticated that it understands English—which is an unnecessary complication. The blueprint/number is the pattern's equivalent of

Self-reproducing Life Pattern

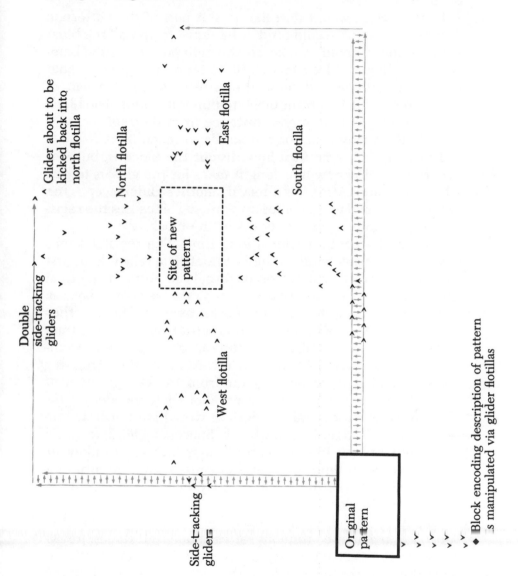

DNA. Like DNA, it would most reasonably use a simple code of its own.

Reproduction would start like this: A part of the pattern, a "supervisory unit" would send out a signal to "read" the blueprint. A glider would scan across the zero position of the blueprint. Finding no block to annihilate there, the glider would complete a circuit and cause the register to shoot a pair of gliders out at the blueprint block to pull it in a unit. The block would surely be an immense distance from the register, so it would take a long time for the flotilla to reach it.

The register has no idea how distant the block is, but it is programmed to know how long it takes for the gliders to pull a block in a unit. After this interval, another glider sweeps the zero position and again comes up empty. This causes the register to fire another pair of gliders at the block.

The process continues for a long time. The register keeps track of how many flotillas it has sent out. The tally is kept first as gliders circulating in a loop (internal memory). If the total grows too large for internal memory, the register may use other external registers (other blocks) as well. Should other external registers be necessary, the register can use more than one. One of these auxiliary registers might represent 1s, another 10s, another 100s, and so on. None of these register blocks would have to be very far from the computer, so it would have quick access to the number they encode.

Eventually, a long train of flotillas would stretch from the register base to the blueprint block. Successive flotillas would have been sent out after increasingly long intervals to allow for increasing travel time—the register must always be sure that the block isn't in position 1,000,000, say, before it sends out the 1,000,001st flotilla.

Finally, the flotillas would reach the block and pull it all the way back in a near-continuous series of one-unit stages. Just as the register is getting ready to shoot out the next flotilla, it scans the zero position and finds the block there. That tells the supervisory unit that the number of flotillas sent out, stored in internal and perhaps external memory, is the number that was in the blueprint register.

Then the supervisory unit switches into action. It feeds the blueprint number—encoded in binary form as gliders and gaps in thin streams—to the main part of the internal computer. The computer mulls over this information and decides how to go about building the structure described. The computer generates the exact pattern of gliders and spaces that will cause the universal constructor to build the structure described in the blueprint. Reading the blueprint register destroys the information in it (as well as the block itself). At some point, the supervisory unit must order the construction of another block (a simple two-glider collision) and move it back out to its original position (with a long, long train of thirty-glider flotillas). Until this is done, the pattern will not be able to reproduce again.

Once the computer has generated the instructions for the universal constructor, the supervisory unit sees that they are conveyed. The universal constructor is a much simpler device than the universal computer. It consists only of side-tracking guns on the pattern's periphery, together with the "wires" (glider streams and gates) connecting them. Given the instructions, the constructor's guns start shooting out gliders.

Most likely, the intended site of the new pattern would lie in one diagonal quadrant of the parent computer. In that case, the side-tracking guns on two sides of the pattern would be activated. The construction process would start with swarms of gliders rising up out of the guns. Initially, the gliders would form two sets of "ladders" ascending from two sides of the pattern. (To simplify the diagram, only two side-tracking "ladders" are shown.) The ladders' "uprights" would lose a glider every time one kicked back a "rung" glider.

Upon reaching the proper height, the gliders in the top rungs would escape. These gliders would double side-track gliders from the central portion of orthogonal ladders. The gliders they kick back would form the two farther glider flotillas needed to construct the new self-replicating pattern. The two near flotillas would escape from the midsections of the ladders by simple side-tracking.

Once the flotillas are complete, the ladders would disappear abruptly. The flotillas would converge, forming eaters, blocks, and finally shuttles. Thus the pattern materializes.

One vital detail is missing. Once the flotillas have been assembled, the supervisory unit of the parent pattern must switch into a new mode of operation. It was interpreting the blueprint as instructions. It must now ignore the instructions and simply interpret the blueprint as a number in short-term memory. It sends a message to the universal constructor to build a block that number of units from the base of the new pattern's blueprint register. Two more gliders are side-tracked out to the proper location and collide to make a block. The new pattern is complete.

EVOLUTION OF LIFE

Conway's vision was this: Start with an infinite Life field, initially in a random state. As the field starts to stabilize, familiar Life objects appear. Most are blocks, blinkers, beehives, gliders, etc.

But even rare objects must turn up somewhere. Search long enough and you can expect to find a paper clip, a heavyweight spaceship, or a shuttle. Rarer yet, two shuttles might happen to form between blocks in just the right way to make a gun. It's not likely, but in an infinite field everything that is possible must happen sometime, someplace.

By the same argument, even such astronomical improbabilities as complete self-reproducing patterns must turn up somewhere in an infinite random field. Once a single self-reproducing pattern forms, it can give birth to another. Its offspring can create still another, and its offspring can make another, and so on—provided only there is enough empty space in which to build.

The most congenial environment for the growth of a colony of self-reproducing patterns would be a space as free of blocks, blinkers, and other Life debris as possible. As mentioned earlier, the stabilized density of a random Life field may be as low as desired by choosing a low initial density. A vanishingly low initial density means that self-reproducing patterns will be all the more unlikely to arise. No matter—we are already reconciled to self-replicators being astronomically rare. The important thing is that once the first self-reproducing pattern forms, it will almost surely have a clear space in which to construct a copy of itself.

As the colony grows, occasionally one member will try to construct a new pattern in a region that already contains a block, blinker, or other object. Sometimes this won't make any difference. The foreign object might happen to lie in between the circuits of the new pattern and not bother anything. Other times, there could be a messy chain reaction that destroys the nascent pattern and that may even send gliders or active regions out to attack other members of the colony. The lower the field density, the less likely such mishaps will be.

For a sparse enough field, new patterns will arise by procreation faster than they are destroyed by foreign objects. Eventually, self-reproducing patterns must tend to fill the Life plane. The most common, most natural large objects in a thin random Life field might—ultimately—be self-reproducing patterns. The fate of any typical region of the plane could be to become populated with self-reproducing patterns, all the offspring of an accidental creation that probably took place in a distant part of the infinite plane.

The Life plane should develop its own ecology. Conway's basic design is probably not the only one possible. Some of the self-reproducing patterns arising in a random field should be better adapted than others.

Some patterns might produce just one offspring and then switch off. Others could be prolific, producing many offspring at different locations relative to themselves.

It would be helpful if patterns could sweep their surroundings clear of Life debris. Perhaps they could side-track gliders to learn what is out there and then shoot out other gliders to eliminate it. Conway did not prove that this is possible, but it seems plausible. As the field density approaches zero, blocks and blinkers predominate over all other Life debris. Both can be eliminated in a glider collision. So a pattern that could deal with just blocks and blinkers would have a tremendous advantage over patterns that forge ahead blind. (For that matter, it is not inconceivable that supersophisticated patterns could deal with the objects in much denser fields. Maybe all infinite random Life fields, and not just ultra-low density ones, become populated with self-replicating patterns.)

Geometry is a barrier to the reproduction of stationary patterns. Offspring crowding around a prolific parent tend to get in the way of further reproduction. Obviously, the parent pattern can't shoot gliders through one of its offspring patterns. Patterns that could move would have a tremendous advantage.

Conway showed that self-replicating Life patterns can pick up and move. This sounds impossible: Guns and eaters are certainly not spaceships. Conway noted, however, that his Life universal constructor was also a universal *de*structor. It can destroy anything that can be annihilated in a glider collision.

Blocks are easily destroyed with a single glider. By shooting a glider at a shuttle in just the right phase, the shuttle can be annihilated too. Four gliders can erase a gun.

Even the eater is not proof against gliders. A glider can creep up from behind and annihilate the eater. It follows, then, that all parts of a self-reproducing pattern can be destroyed by a suitable swarm of gliders.

Conway realized that a self-reproducing pattern could send out the gliders to replicate itself—having them converge on some distant location—and then assemble the glider flotillas that will destroy itself, pointing them at itself. The old pattern would disappear, but it would reappear exactly at another location. That makes it a spaceship.

Only a few design changes are necessary for self-replicating patterns to move in this fashion. A pattern should erase itself completely, not leaving any stray gliders. To ensure this, the

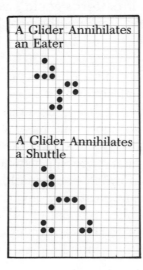

A Glider Annihilates an Eater

A Glider Annihilates a Shuttle

guns should be shot down first, leaving the eaters in place to consume all the gliders still circulating through the pattern.

Several of the constructions diagramed in this chapter, such as the thin gun and the logic gates, do not use eaters where a coded stream meets a full stream in a vanish reaction. Should the full-stream gun be destroyed first, the coded stream will escape as stray gliders. Consequently, a pattern that may need to annihilate itself should have an eater for every gun, even where they are not otherwise needed.

Furthermore, each gun must be located so that gliders arriving from outside have a clear path to it—so that it is not necessary to annihilate any eaters in order to annihilate all guns.

That leaves only the memory blocks, which are normally outside the pattern. The pattern produces the flotillas that will destroy its guns and eaters only. It sends the flotillas far out from itself so that it will have time to take care of the memory blocks before the flotillas arrive. It then simply pulls all the blocks all the way in. The zero test automatically destroys the memory blocks.

Then the leading gliders in the flotillas destroy the guns; the remaining eaters consume all of the pattern's remaining gliders; and the tail ends of the flotillas wipe out the eaters.

A self-replicating Life pattern that moves might benefit from sensory and nervous systems to deal with new environments. Double side-tracking can send a glider out and bring it back. Gliders that fail to return would indicate an obstacle ahead. Thus a pattern could have "eyes" or "antennae" of sorts.

Presumably, the long-term evolution of an infinite Life field would proceed somewhat like biologists imagine the evolution of life on the early earth proceeded.

First the Life plane would become populated with colonies of first-generation self-replicating patterns. Each colony would derive from a single seed pattern that arose purely by chance. The odds against a pattern arising by chance grow even steeper for more complex patterns, so practically all of the first-generation patterns would possess only the minimum level of organization necessary to reproduce. They would be the Life version of Von Neumann's machine—a mindless self-replicator.

From time to time, mutations would occur. A pattern might be constructed in a site containing other Life objects. A stray glider might wander into a pattern's works. Most such mutations would be harmful. But rare cases would produce patterns better able to survive than their neighbors. Natural selection would favor the mutated patterns. The species of self-reproducing patterns in a given region would gradually change through mutation, extinction, and migration from distant regions.

The existence of universal computers in Life implies that the nervous systems of self-reproducing patterns can be arbitrarily complex. Eventually, "intelligent" species might evolve from the same sort of selective pressures that operated on earth.

If nontrivial self-reproduction is used as the criterion of life, then self-reproducing Life patterns would be alive. This is not to say that they would simulate life as any television image might, but that they would be literally alive by virtue of encoding and manipulating information about their own makeup. The simplest self-reproducing Life patterns would be alive in a sense that a virus is not.

HOW BIG ARE SELF-REPRODUCING LIFE PATTERNS?

It must be stressed again that "living" Life patterns would be huge. And the random field required for spontaneous generation of self-reproducing patterns would be vast beyond imagining. Let's estimate.

We're interested mainly in the minimum size of self-reproducing Life patterns. A self-reproducing pattern may be any size above the minimum. Conway did not try to guess the size of such patterns.

There is Von Neumann's estimate of 200,000 cells for his cellular self-reproducing machine. But that was in Von Neumann's 29-state space, not Life. A self-reproducing Life pattern would be much, much larger than 200,000 pixels. Two states must do the work of Von Neumann's 29. Von Neumann's space was tailor-made for self-reproduction; Life is not.

There is no simple correspondence between Von Neumann's 29 states and the components of a Life self-reproducing pattern. So we will have to settle for a very, very rough estimate.

In both the Von Neumann and the Conway automaton, the most fundamental structural unit is the bit. Here, *bit* means the section of "wire" or position in a glider stream that may or may not contain a pulse/glider. It is that area of the pattern that encodes one bit of information in the pattern's computer circuitry.

In Von Neumann's model, a single cell encodes a bit. The effective bits of Life computers are much larger. Not only the five pixels that may contain the glider but all the surrounding empty pixels must be counted.

The gliders in a full-strength, period-30 stream are 7½ pixels apart diagonally. Think of each glider position as being in the center of a 7½-by-7½-pixel diagonal square. The area of this square comes to 112.5 pixels (twice 7½ times 7½ because the sides are measured diagonally rather than orthogonally).

The basic, information-carrying streams in a self-reproducing Life pattern will not be full strength. They will be thinned. No one knows what thinning factor will suffice. Conway guessed 1000. Assume, then, that the streams are thinned 1000 times. Then there are 7500 pixels between glider positions in streams. We must allow a square measuring 7500-by-7500 pixels diagonally for each glider or space. The area of this square is 112,500,000 pixels. This might as well be rounded to an even 100 million, or 10^8.

Von Neumann's machine is 200,000 times larger than a single bit. To an order of one or two magnitude, this ratio probably holds for minimal self-reproducing Life patterns, too. That means that the size of a self-reproducing Life pattern should be about 200,000 times 100 million pixels. This comes to 20 trillion or 2×10^{13} pixels. There is little point in keeping the 2 in view of the uncertainties in the estimate, so let's say 10^{13} pixels.

That's big. The text mode of a typical home computer has about 1000 to 2000 character positions. A high-resolution graphics mode might display 100,000 pixels.

Displaying a 10^{13}-pixel pattern would require a video screen about 3 million pixels across at least. Assume the pixels are 1 millimeter square (which is very high resolution by the standards of home computers). Then the screen would have to be 3 kilometers (about 2 miles) across. It would have an area about six times that of Monaco.

Perspective would shrink the pixels of a self-reproducing pattern to invisibility. If you got far enough away from the screen so that the entire pattern was comfortably in view, the pixels (and even the gliders, eaters, and guns) would be too tiny to make out. A self-reproducing pattern would be a hazy glow, like a galaxy.

It is practical only to set a lower limit on the chances of a self-reproducing pattern forming in a random field. For simplicity, assume a field at 50 percent density. Take a region of 10^{13} pixels. There ought to be at least one way of fitting a self-reproducing pattern into that region. Every pixel in Life has two possible states. The total number of configurations for a 10^{13} pixel region is $2 \times 2 \times 2 \times 2 \times 2 \times \ldots \times 2$, where there are 10^{13} factors of 2. This can be expressed as $2^{10^{13}}$. Clearly the chances of the region containing a self-reproducing pattern can be no worse than 1 in $2^{10^{13}}$, even in the initial random state.

We can also place an upper limit on the minimum size of field that would be required for spontaneous formation of self-reproducing patterns. Assume that 1 in $2^{10^{13}}$ is the chance that a typical 10^{13}-pixel region contains a self-reproducing pattern. The chances of finding a self-reproducing pattern are greater, of course, if more than one 10^{13}-pixel region is searched. On the average, you could expect to find a self-reproducing pattern for every $2^{10^{13}}$ 10^{13}-pixel regions searched. That amounts to $2^{10^{13}}$ times 10^{13} pixels.

When 1 is added to a trillion, the sum is still approximately a trillion. The same sort of thing happens in exponential notation even when numbers are multiplied. $2^{10^{13}}$ can be reexpressed approximately as $10^{10^{12}}$. $10^{10^{12}}$ times 10^{13} is a number that still rounds to $10^{10^{12}}$ (or the original $2^{10^{13}}$).

This is only an upper limit. It seems pretty certain that self-reproducing patterns would form in a random field of $10^{10^{12}}$ pixels. Probably they would form in fields much, much smaller. Unfortunately, there is no easy way of telling how much smaller. $10^{10^{12}}$ is not a very tight limit. A video screen as wide as the observable universe would not have nearly a googol (10^{10^2}) pixels.

Conway's fantasy of living Life patterns forming from video snow will almost certainly never be realized on a computer. It is barely conceivable that someone could someday go to the trouble of designing a self-reproducing Life pattern in full detail. It is possible that computer technology could advance exponentially to the point where such a pattern could be tracked through its reproduction. Spontaneous creation of self-reproducing patterns is another matter, however. Not even the rosiest predictions for future computer technology are likely to make any difference. Although $10^{10^{12}}$ pixels may not be necessary, it seems highly unlikely that a computer/video screen any smaller than (as a guess) the solar system could suffice. Self-reproducing patterns must arise on a big enough screen, but no screen ever likely to be built will be big enough.

·13·

THE RECURSIVE UNIVERSE

Over the Labor Day weekend of 1949, the ENIAC computer calculated pi to 2,035 decimal places. Von Neumann became briefly interested in the extended expansions of pi and other mathematical constants. He subjected them to statistical analysis, looking for subtle patterns.

Pi, or any recursively defined entity, is the information-theory equivalent of a perpetual-motion machine. It takes a certain amount of information to build the machine—that is, to describe the recursive rules unambiguously. Then the recursion grinds out information endlessly. It is something for nothing, or rather everything for almost nothing.

This aspect of recursive definitions defies common sense. We are so inured to the first law of thermodynamics that it seems, by analogy, impossible for a mathematical procedure to produce a surplus of information. Pi can be described unequivocally in less than ten words: "the ratio of a circle's circumference to its diameter." If necessary, the definition can include definitions of "circle," "circumference," and "diameter" and a few of the explicitly recursive formulas for pi. That short definition allows anyone to calculate the endless train of digits in the expansion of pi. Since the definition of pi implicitly specifies a digit for each of an infinity of decimal places, it embodies an infinity of information.

Imagine a super-fast computer hooked up to a television set. The computer calculates the digits of pi and transmits them to the television screen. The screen uses the digits to specify the states of successive (ten-state) pixels. Thus the simple definition of pi specifies a TV "show" that will go on forever.

Von Neumann's analysis found no evidence for patterns in pi beyond what one would expect in a perfectly random string of digits. By the same token, pi's evident randomness virtually ensures that all possible patterns of digits occur somewhere in the expansion.

It is now known, for instance, that a string of six consecutive 9s occurs. This would flash by as a monocolor spike on the TV screen. By and large, pi's expansion would produce video snow, but any conceivable TV image must occur somewhere in pi. If you watched long enough, you would even see pi run through every possible TV show. And yet everything you would see is somehow latent in "the ratio of a circle's circumference to its diameter."

Past generations were most concerned with the material side of creation: How could something come from nothing? Cosmologists now recognize another, informational side of creation. Creation of a world entails above all information. The state of everything—everywhere—at every time—must be defined. The most economical way to specify such information is through a complexity-generating recursion of physical law.

Physics deals with changes at successive instants; in that trivial sense it has always involved recursion. The theists of Kant's, Laplace's, and Leibniz's times supposed that a Creator had tailored an initial state so as to lead to human life and the other complexities of the world. Then the Creator sat back and watched His creation play out its fate. All the complexity of the world was embodied in that initial state. The laws of physics were just a sort of mainspring to keep things moving.

The cosmology of the past decades has increasingly devalued the importance of the initial microstates. It has become a guiding principle of cosmology to invoke no special conditions in the early universe. The world is assumed to have started in a state of thermodynamic equilibrium, and the exact microstate is assumed unimportant to the general trend of cosmic evolution. In such a model, physical laws are foremost.

There is better reason than ever for believing that the structure and complexity of our world is inherent in our physical laws and not in some special, unknowable microstate. The universe is a recursively defined geometric object. Von Neumann's automatons have relevance for cosmology as well as biology. That a machine can manufacture ever more complex descendants is much more than a theorem of electrical engineering. On a deeper level, Von Neumann was really talking about what is possible within the context of physical law. He showed that structure can grow richer under physical law (and nothing else). The processes that led to the formation of galaxies are nothing like the reproduction of Von Neumann's machines, but the object lesson is clear: Complexity is self-generating. The diversity of our world is understandable because it is possible to design imaginary self-consistent worlds potentially as complex as our own.

This is no mere restatement of common sense. Everyone daydreams alternate worlds, but the imagination soon tires of filling in details. Ulam, Von Neumann, and Conway showed that a few recursive rules can paint in all the details. Creation can be simple.

LIFE FOR HOME COMPUTERS

A number of software firms market Life programs on magnetic media. Commercial software is convenient, fast, and colorful. A color monitor and adaptor are often required.

Life is easily programmed into BASIC. The only drawback is speed. A single generation may take 15 seconds or more with a modest field size. However, it is possible to buy a BASIC compiler disk for many computers. A BASIC compiler translates a BASIC program into object code for much greater speed of execution.

Assembly-language programs are as fast as the commercial software. Gliders move, blinkers blink, honey farms bloom. But assembly language varies from model to model: An IBM PC assembly-language program won't run on an Apple. An assembly-language program must be "assembled" using an assembler disk that must be purchased. Unless you plan to use the assembler disk for other programs, you're probably better off buying ready-to-use Life software or a BASIC compiler.

The Life plane is infinite; computers are not. There are two customary strategies for dealing with the finitude of computer displays and memories. In both cases, the computer tracks the evolution of a finite rectangular region only.

A "bounded" program acts as if the pixels beyond this rectangular region are forever empty. Pixels on the edge have only five neighbors, in effect, for the three imaginary neighbors beyond the edge can never switch on. (The four corner pixels of the rectangular region have just three neighbors.)

Once a pattern touches the edge, its evolution is almost

always different from what it would be on an infinite display. Not only is growth beyond the edge halted, but also the evolution within the rectangular tracking regions may be different —since this may well be influenced by growth that would extend beyond the edge.

Gliders that hit the edge are truncated to a four-pixel form that becomes a block in two generations. A glider gun can be safely contained in a bounded field because its odd-numbered gliders turn into blocks when they hit the edge and its even-numbered gliders annihilate the blocks.

A lightweight spaceship hits the edge and bounces back as a glider. A mediumweight spaceship forms *half* a pulsar, and a heavyweight spaceship vanishes. In some ways, the edge of a bounded field is like a mirror. Half a pulsar clinging to the edge is a stable oscillator because the edge prevents growth like the missing quadrants would. George D. Collins, Jr., discovered the tumbler, in halved form, at the edge of a bounded field.

A "toroidal" display wraps around: Things that go off one end of the screen reappear at the opposite edge. The top row of pixels has the bottom row for neighbors, and vice versa; the pixels in the leftmost column are treated as neighbors of the pixels in the rightmost column.

Fast toroidal displays are most convenient for demonstrating glider and spaceship motion. Several gliders or spaceships will swim around a toroidal screen like fish in an aquarium. Ultimately, gliders or spaceships will collide with unpredictable results.

A glider gun's stream may wrap around into a number of parallel diagonal lanes. Eventually, the stream will take aim on the gun itself and all will dissolve into chaos. If you don't want the gliders to wrap around, a properly positioned eater will absorb them.

Toroidal displays percolate longer than bounded displays in random-field experiments. The toroidal field is a better model of an infinite field because active regions always interact with other Life objects rather than an unyielding edge.

Sophisticated Life programs may have features making it easier to enter, store, and modify patterns. "Step execution" is a handy feature for fast programs. Successive generations are displayed at each touch of a key. The display may be frozen for study. In various color versions of Life, two interacting patterns may be assigned different colors. Surviving pixels do not change color. The color of birth pixels is determined by majority vote of the three parents. Slightly more complicated rules allow for additional colors.

Below are a simple BASIC program adaptable (with modifications where necessary) to any small computer and an IBM PC assembly-language program.

BASIC LIFE

This is a program for a 22-by-40 pixel bounded display. If your computer has a larger display, the program is easily modified to fill the screen. Asterisks represent on pixels. You will probably want to replace the asterisks (line 200) with rectangular or circular characters for a clearer display. Asterisks are used here because the codes for designating these characters vary from model to model. Lines 75 and 900 to 970 are optional. They use BASIC's random-number feature to generate a random field.

Initial patterns are input by drawing them row by row. Use the zero key to input on pixels and the period key to represent off pixels. After the program prompts "Enter a pattern (type END when finished):," enter

```
. . . . . 0 .      (carriage return)
. . . . . . 0      (carriage return)
0 . . . . . 0      (carriage return)
. 0 0 0 0 0 0      (carriage return)
END                (carriage return)
```

to input a heavyweight spaceship traveling to the right. It is necessary that the pattern of zeros and periods be rectangular —each row must be filled out to the same number of characters with periods—and that END be typed in capital letters. The pattern will automatically be centered in the field. Patterns may be positioned off-center by inputting extra periods on one side of the pattern.

When lines 75 and 900 to 970 are included, a random field

can be generated by typing RANDOM (carriage return) after the prompt. (Again, RANDOM must be in capital letters.) The program will ask for a random number seed (enter any number in the range specified) and a density for the random field (a decimal fraction between 0 and 1; try .3). Random fields run slowly because they occupy the entire display.

The program displays the generation at the bottom of the screen.

```
10 CLS
20 PRINT "LIFE"
30 PRINT: PRINT: PRINT
40 DIM A(40,22), B$(22)
50 H=1
60 PRINT "Enter a pattern (type END when finished):"
70 INPUT B$(H)
75 IF B$(H)="RANDOM" THEN GOTO 900
80 IF B$(H)="END" THEN H=H-1: GOTO 120
90 IF LEN(B$(H))>W THEN W=LEN(B$(H))
100 H=H+1
110 GOTO 70
120 XMIN=INT(20-W/2): YMIN=INT(11-H/2): XMAX=XMIN+W: YMAX=YMIN+H
130 FOR Y=1 TO H: FOR X=1 TO W
140 IF MID$(B$(Y),X,1)="." THEN GOTO 160
150 A(X+XMIN,Y+YMIN)=1
160 NEXT X: NEXT Y: CLS
170 RIGHT=1: LEFT=40: TOP=22: BOTTOM=1
180 FOR Y=YMIN TO YMAX: LOCATE Y,XMIN: FOR X=XMIN TO XMAX
190 IF A(X,Y)=11 THEN A(X,Y)=1: IF A(X,Y)=10 THEN A(X,Y)=0
200 IF A(X,Y)=1 THEN PRINT "*";: GOTO 220
210 PRINT " ";: GOTO 260
220 IF X>RIGHT THEN RIGHT=X
230 IF X<LEFT THEN LEFT=X
240 IF Y<TOP THEN TOP=Y
250 IF Y>BOTTOM THEN BOTTOM=Y
260 NEXT X: NEXT Y
270 FOR Y=YMAX+1 TO 22: PRINT: NEXT Y
280 PRINT "TIME"; T
290 XMIN=LEFT: YMIN=TOP: XMAX=RIGHT: YMAX=BOTTOM
300 IF XMIN<3 THEN XMIN=3
310 IF YMIN<3 THEN YMIN=3
320 IF XMAX>38 THEN XMAX=38
330 IF YMAX>20 THEN YMAX=20
340 T=T+1: XMIN=XMIN-1: YMIN=YMIN-1: XMAX=XMAX+1: YMAX=YMAX+1
350 FOR Y=YMIN TO YMAX: FOR X=XMIN TO XMAX
360 NBRS=A(X-1,Y-1)+A(X,Y-1)+A(X+1,Y-1)+A(X-1,Y)
370 NBRS=INT(NBRS/10)
380 NBRS=NBRS+A(X+1,Y)+A(X-1,Y+1)+A(X,Y+1)+A(X+1,Y+1)
390 NBRS=NBRS-10*INT(NBRS/10)
400 A(X,Y)=A(X,Y)-10*INT(A(X,Y)/10)
410 A(X,Y)=10*A(X,Y)
420 IF NBRS=3 THEN GOTO 450
430 IF NBRS<>2 THEN GOTO 460
440 IF A(X,Y)<>10 THEN GOTO 460
450 A(X,Y)=A(X,Y)+1
460 NEXT X:NEXT Y
470 GOTO 180
480 END
900 RANDOMIZE
910 PRINT "Enter density:"
```

```
920 INPUT D
930 FOR Y=2 TO 21: FOR X=2 TO 39
940 IF RND(X+Y)<D THEN A(X,Y)=1
950 NEXT X: NEXT Y: CLS
960 XMIN=3: YMIN=3: XMAX=38: YMAX=20
970 GOTO 180
```

IBM PERSONAL COMPUTER ASSEMBLY-LANGUAGE LIFE

This is a much more powerful program for a 50-by-80-pixel toroidal display. The field is large enough to show the "lifelike" or "amoeboid" appearance of large active regions. About 1.1 generations are displayed per second. The program may be modified to run much faster with a smaller field, but for most purposes, the larger field is likely to be more useful than the faster speed. A cursor facilitates entry of patterns. The display may be halted, modified, and started over any numer of times. To run this program, you will need an IBM PC with monochrome display (or compatible system), any version of the disk-operating system (DOS), and the IBM Assembler or Macro Assembler.

Enter the program with the EDLIN text editor (on the DOS disk). Then assemble it with the IBM Assembler program and link it with the LINK utility (on the DOS disk) as explained in the DOS and Assembler instructions. Once this is done, the program can be run simply by entering the file name you assign (say, LIFE) after the DOS prompt.

The program displays a screen of instructions and then blanks out the display. A cursor appears in the center of the screen. The numeric keypad (the extra set of number keys on the right of the keyboard) is active; all other keys except the ESCAPE key will be ignored. The INSERT key switches pixels on; DELETE erases. The cursor-arrow keys work as expected, except that it may take two strokes of the cursor-up or cursor-down key to move the cursor. (There are 25 allowed vertical positions for the cursor versus 50 for the program's display.

Pushing the cursor-up key shifts the action of a subsequent INSERT or DELETE key up one pixel, though it might take another stroke to move the cursor up.) Pressing the ESCAPE key starts the Life display.

Subsequently, hitting any key (other than ESCAPE) will halt the display and restore the cursor for editing the Life field. Hitting ESCAPE during a run will return you directly to the operating system.

A random field may be simulated by entering several long rows or columns of on pixels, irregularly spaced. Their interaction will soon produce chaotic active regions occupying most of the screen.

This program will run on many IBM-compatible systems. Make sure that the variable "VIDEO" contains the correct starting address of video memory.

```
;IBM ASSEMBLER LANGUAGE LIFE
;50 BY 80 PIXEL FIELD
;-------------------------------------------------------------
STACK     SEGMENT PARA      STACK     'STACK'
          DB        256 DUP(0)
STACK     ENDS
;-------------------------------------------------------------
DATA      SEGMENT PARA      PUBLIC   'DATA'
VIDEO     =0B000H ;IMPORTANT NOTE ON COMPATIBILITY: THE VARIABLE
                  ;"VIDEO" MUST BE SET TO THE STARTING ADDRESS OF
                  ;VIDEO MEMORY--0B000H FOR THE IBM PC. THIS PROGRAM
                  ;WILL RUN ON MANY IBM-COMPATIBLE COMPUTERS IF THE
                  ;CORRECT VIDEO ADDRESS IS SUBSTITUTED IN THIS LINE.
                  ;IT'S 0B800H FOR A COMPAQ.
ARRAY     DB        4000 DUP(0)
COMPLETE            DB        0
BANNER    DB        'Game of Life'
          DB        13,10,10,10
          DB        'Copyright 1984 by William Poundstone'
          DB        13,10,10
          DB        'Life is an abstract computer animation invented by British'
          DB        13,10
          DB        'mathematician John Horton Conway.  To start the game, use'
          DB        13,10
          DB        'the numeric keypad to enter a pattern of square blocks of'
          DB        13,10
          DB        'light ("pixels").  The cursor-movement keys shift the cursor'
          DB        13,10
          DB        'around the screen.  Pressing the INSERT key will switch a'
          DB        13,10
          DB        'pixel on at the cursor position.  If you make a mistake, use'
          DB        13,10
          DB        'the DELETE key to switch pixels off.  When the pattern is'
          DB        13,10
          DB        'complete, press the ESCAPE key to start the animation.'
          DB        13,10,10
          DB        'Hitting any letter key during a run freezes the action and'
          DB        13,10
          DB        'restores the cursor.  You may edit the pattern or simply'
          DB        13,10
          DB        'press the ESCAPE key to resume.  Hitting the ESCAPE key'
          DB        13,10
          DB        'during a run returns you to the operating system.'
          DB        13,10,10
          DB        'Hit any key to begin.',',','$'
DATA      ENDS
;-------------------------------------------------------------
CODE      SEGMENT PARA      PUBLIC   'CODE'
START     PROC      FAR
```

```
;PRELIMINARY STATEMENTS REQUIRED IN IBM ASSEMBLER LANGUAGE
        ASSUME  CS:CODE
        PUSH    DS
        MOV     AX,0
        PUSH    AX
        MOV     AX,DATA
        MOV     DS,AX
        ASSUME  DS:DATA
;---------------------------------------------------------------------
;CLEAR SCREEN BY SCROLLING
        MOV     AX,0600H
        MOV     BH,07   ;NORMAL ATTRIBUTE
        MOV     CX,00   ;BOUNDS FOR . . .
        MOV     DX,184FH;. . . FULL SCREEN
        INT     10H     ;CLEAR SCREEN
;---------------------------------------------------------------------
;DISPLAY BANNER
        MOV     DX,OFFSET BANNER;LOAD ADDRESS
        MOV     AH,9    ;DISPLAY REQUEST
        INT     21H     ;LET DOS DISPLAY INSTRUCTIONS
;---------------------------------------------------------------------
;WAIT FOR USER TO READ INSTRUCTIONS AND HIT ANY KEY
        MOV     AH,0    ;KEYBOARD INPUT REQUEST
        INT     16H     ;LET BIOS WAIT FOR KEYBOARD INPUT
;THE CHARACTER ENTERED IS PUT IN AL BUT WE DON'T NEED IT
;---------------------------------------------------------------------
;CLEAR SCREEN AGAIN
        MOV     AX,0600H
        MOV     BH,07   ;NORMAL ATTRIBUTE
        MOV     CX,00   ;BOUNDS FOR . . .
        MOV     DX,184FH;. . . FULL SCREEN
        INT     10H     ;CLEAR SCREEN
;---------------------------------------------------------------------
;CENTER CURSOR IN MIDDLE OF SCREEN
;BH WILL BE CURSOR ROW AND BL WILL BE COLUMN
;BUT DH AND DL WILL BE ROW AND COLUMN OF 50 BY 80 FIELD
CENTER: MOV     BX,0C28H        ;COORDINATES FOR CENTER
        MOV     DX,1928H        ;COORDINATES FOR CENTER ALSO
;SET ES TO VIDEO MEMORY
        MOV     AX,VIDEO
        MOV     ES,AX
;---------------------------------------------------------------------
;DISPLAY CURSOR AT CURRENT POSITION
INPUT:  MOV     AX,0    ;CALCULATE PIXEL POSITION . . .
        PUSH    BX
        PUSH    DX
        MOV     AL,DH
        MOV     BL,DL
        MOV     DL,80
        MUL     DL
        MOV     BH,0
        ADD     AX,BX
        POP     DX
        POP     BX
        MOV     BP,AX   ; . . . PUT IT IN BP
;SET CURSOR ACCORDING TO BH AND BL
        PUSH    DX
        MOV     DX,BX
        MOV     AH,02   ;SET CURSOR REQUEST
        PUSH    BX
        MOV     BH,00   ;PAGE 0
        INT     10H     ;LET BIOS SET CURSOR (IT USES CONTENTS OF DX,
                        ;HENCE THE SWITCH)
        POP     BX      ;RESTORE CURSOR ROW AND COLUMN IN BH AND BL
        POP     DX      ;RESTORE PIXEL ROW AND COLUMN IN DH AND DL
;---------------------------------------------------------------------
;KEYBOARD ROUTINE DETECTS ENTRIES AND TAKES APPROPRIATE ACTION.
;THE CURSOR-ARROW KEYS MOVE THE CURSOR.  INSERT SWITCHES PIXELS
;ON.   DELETE ERASES.  THE ESCAPE KEY STARTS RUN.  ALL OTHER KEYS
;ARE IGNORED.
        MOV     AX,0    ;INPUT CHARACTER GOES INTO AL
        MOV     SI,OFFSET ARRAY ;SI WILL BE MEMORY LOCATION OF CELL
        MOV     AH,7    ;KEYBOARD REQUEST CODE
        INT     21H     ;LET DOS READ KEYBOARD INPUT
ESCAPE: CMP     AL,27   ;ESCAPE KEY TO TERMINATE?
        JNZ     KEYPAD  ;ESCAPE IS ONLY ACTIVE NON-KEYPAD KEY
        INC     COMPLETE;THIS VARIABLE SIGNIFIES THAT PATTERN IS
                        ;COMPLETE AT THE BRANCH BETWEEN THE
                        ;DISPLAY AND MAIN ROUTINES
```

```
; TURN CURSOR OFF IF PATTERN IS COMPLETE
          MOV     AH,01     ;SET CURSOR TYPE REQUEST
          MOV     CH,2BH    ;NO CURSOR DISPLAYED
          INT     10H       ;LET BIOS SWITCH OFF CURSOR
          JMP     DISPLAY
KEYPAD:   CMP     AL,0      ;A NULL CHARACTER IS RETURNED FOR KEYPAD KEYS
          JNZ     INPUT     ;IGNORE NONFUNCTIONAL KEYS
          INT     21H       ;READ AGAIN TO GET KEYPAD KEY
RIGHT:    CMP     AL,77     ;RIGHT KEY?
          JNZ     DOWN
          INC     BL        ;INCREMENT CURSOR COLUMN
          INC     DL        ;INCREMENT PIXEL COLUMN
          JMP     INPUT
DOWN:     CMP     AL,80     ;DOWN KEY?
          JNZ     LEFT
          INC     DH        ;INCREMENT PIXEL ROW
          PUSH    DX
          SHR     DH,1      ;IS PIXEL ROW NOW EVEN?
          JC      SKIPD     ;IF NOT, BRANCH
          INC     BH        ;IF SO, INCREMENT CURSOR ROW
SKIPD:    POP     DX
          JMP     INPUT
LEFT:     CMP     AL,75     ;LEFT KEY?
          JNZ     UP
          DEC     DL        ;DECREMENT PIXEL COLUMN
          DEC     BL        ;DECREMENT CURSOR COLUMN
          JMP     INPUT
UP:       CMP     AL,72     ;UP KEY?
          JNZ     DELETE
          DEC     DH        ;DECREMENT PIXEL ROW
          PUSH    DX
          SHR     DH,1      ;IS PIXEL ROW NOW ODD?
          JNC     SKIPU     ;IF NOT, BRANCH
          DEC     BH        ;DECREMENT CURSOR ROW
SKIPU:    POP     DX
          JMP     INPUT
DELETE:   CMP     AL,83     ;DELETE KEY?
          JNZ     INSERT
          MOV     AX,BP
          ADD     SI,AX
          MOV     AX,0
          MOV     [SI],AL   ;ERASE PIXEL
          JMP     DISPLAY
INSERT:   CMP     AL,82     ;INSERT KEY
          JZ      SWITCH
          JMP     INPUT     ;IGNORE ANY NONFUNCTIONAL KEY
SWITCH:   MOV     AX,BP
          ADD     SI,AX
          MOV     AX,1
          MOV     [SI],AL   ;STORE A 1 FOR ON CELL
;----------------------------------------------------------------
;VIDEO DISPLAY
;GRAPHIC BLOCK CHARACTERS (219, 220, AND 223) ARE USED TO
;GET TWO PIXELS OUT OF EACH CHARACTER POSITION
DISPLAY:MOV       SI,OFFSET ARRAY ;SI POINTS TO BASE OF MEMORY
          MOV     BP,0      ;IN THIS ROUTINE, BP IS A COUNTER THAT TELLS
                           ;WHEN THE PROGRAM COMES TO THE END OF A ROW
          MOV     CX,2000
          MOV     DI,0
          PUSH    BX        ;SAVE CURSOR . . .
          PUSH    DX        ;. . . AND PIXEL LOCATION
SCREEN:   MOV     DL,[SI]
          MOV     AL,' '    ;DEFAULT OF BLANK CHARACTER . . .
          AND     DL,01H    ; . . . UNLESS DL'S LOWER NIBBLE IS 1
          JZ      OFFTOP
          MOV     AL,223
          MOV     DL,[SI+80]
          AND     DL,01H
          JZ      FILL
          MOV     AL,219
          JMP     FILL
OFFTOP:   MOV     DL,[SI+80]
          AND     DL,01H
          JZ      FILL
          MOV     AL,220
FILL:     MOV     ES:[DI],AL      ;PUT CHARACTER IN ADAPTER MEMORY
          ADD     DI,2      ;SKIP OVER ATTRIBUTE BYTE
          INC     BP
          CMP     BP,80
```

```
        JNZ      USUAL
        ADD      SI,80
        SUB      BP,80
USUAL:  INC      SI
        LOOP     SCREEN      ;DO NEXT CHARACTER POSITION
        POP      DX
        POP      BX          ;RECOVER CURSOR, PIXEL LOCATIONS
;-------------------------------------------------------------------
;BRANCH BACK TO INPUT ROUTINE IF USER IS STILL IS STILL ENTERING
;PATTERN
        CMP      COMPLETE,0    ;IT WILL BE 1 IN ANIMATION MODE
        JNZ      INTERRUPT
        JMP      INPUT
;-------------------------------------------------------------------
;CHECK TO SEE IF ANY KEY HAS BEEN STRUCK.  IF SO, HALT EXECUTION.
INTERRUPT:MOV    AH,1
        INT      16H
        JNZ      RESTART ;IF KEY HAS BEEN STRUCK, LEAVE LOOP
        JMP      MAIN    ;OTHERWISE CONTINUE GENERATIONS
RESTART:JMP      NEW
;-------------------------------------------------------------------
```

▶ Step execution Insert for
IBM Assembler Program

```
INTERRUPT: MOV AX,0
INT 16H; WAIT FOR KEY TO BE STRUCK
CMP AL,46; IS IT A PERIOD TO TERMINATE?
JZ EXIT
```

```
;MAIN ROUTINE APPLIES THE RULES OF LIFE TO "ARRAY" IN DS
;EACH BYTE IN "ARRAY" REPRESENTS ONE PIXEL THUS: 000?000?.
;HERE THE RIGHT "?" IS A 0 OR 1 DEPENDING ON CURRENT STATE OF
;PIXEL.  THE LEFT "?" REPRESENTS THE PREVIOUS STATE.  THE PROGRAM
;USES A POPULAR LIFE ALGORITHM--PERFORM A LOGICAL "OR" ON THE
;SUM OF THE NEIGHBORS AND 1.  ONLY IF THE RESULT IS 3 WILL THE
;PIXEL BE ON IN THE NEXT GENERATION.
MAIN:   MOV      BX,0      ;BL WILL BE SUM OF NEIGHBORS
        MOV      DX,0      ;DH WILL HOLD ROW AND DL WILL HOLD COLUMN
        MOV      CX,4000
        MOV      SI,OFFSET ARRAY ;SI WILL BE DISPLAY POSITION
POSITION:        MOV      AX,80    ;CALCULATE DISPLAY POSITION
                                   ;FROM DH AND DL . . .
        PUSH     DX
        MUL      DH
        MOV      DH,0
        ADD      AX,DX
        MOV      BP,AX
        POP      DX        ; . . . AND PUT IT IN BP, RESTORING DH AND DL
WEST:   CMP      BP,1
        JL       EDGEW
        ADD      BL,[SI-1]
        JMP      NORTHEAST
EDGEW:  ADD      BL,[SI+3999]
NORTHEAST:       CMP      BP,79
        JL       EDGENE
        ADD      BL,[SI-79]
        JMP      NORTH
EDGENE: ADD      BL,[SI+3921]
NORTH:  CMP      BP,80
        JL       EDGEN
        ADD      BL,[SI-80]
        JMP      NORTHWEST
EDGEN:  ADD      BL,[SI+3920]
NORTHWEST:       CMP      BP,81
        JL       EDGENW
        ADD      BL,[SI-81]
        JMP      DIVIDE
EDGENW: ADD      BL,[SI+3919]
DIVIDE: PUSH     CX        ;SAVE COUNT WHILE USING CX FOR DIVISION
        MOV      CX,4
        SHR      BL,CL     ;RECOVER PREVIOUS STATES OF NEIGHBORS ALREADY SERVI
        POP      CX        ;RECOVER COUNT
EAST:   CMP      BP,3998
        JG       EDGEE
        ADD      BL,[SI+1]
        JMP      SOUTHWEST
EDGEE:  ADD      BL,[SI-3999]
SOUTHWEST:       CMP      BP,3920
        JG       EDGESW
        ADD      BL,[SI+79]
        JMP      SOUTH
```

```
EDGESW: ADD      BL,[SI-3921]
SOUTH:  CMP      BP,3919
        JG       EDGES
        ADD      BL,[SI+80]
        JMP      SOUTHEAST
EDGES:  ADD      BL,[SI-3920]
SOUTHEAST:  CMP      BP,3918
        JG       EDGESE
        ADD      BL,[SI+81]
        JMP      BIRTH
EDGESE: ADD      BL,[SI-3919]
BIRTH:  OR       BL,[SI]
        MOV      AL,[SI] ;GET READY TO MULTIPLY OLD VALUE BY 16
        PUSH     CX
        MOV      CX,4
        SHL      AL,CL    ;MULTIPLY
        POP      CX
        AND      BL,07H   ;IS BL MOD 8 EQUAL TO 3?
        CMP      BL,3
        JNZ      READY
        ADD      AL,1     ;A LIVE CELL IN THE NEW GENERATION
READY:  MOV      [SI],AL  ;PUT VALUE BACK IN ARRAY
        INC      SI       ;MOVE TO NEXT ARRAY LOCATION
        MOV      BX,0     ;CLEAR SUM
        INC      DL       ;NEXT COLUMN POSITION
        CMP      DL,80    ;OFF EDGE?
        JNZ      LOOP
        INC      DH       ;INCREMENT ROW NUMBER
        MOV      DL,0     ;RESET COLUMN NUMBER
LOOP:   DEC      CX
        CMP      CX,0
        JZ       CYCLE
        JMP      POSITION ;DO NEXT PIXEL
;----------------------------------------------------------------
;DISPLAY NEW GENERATION
CYCLE:  JMP      DISPLAY
;----------------------------------------------------------------
;THIS ROUTINE TURNS CURSOR BACK ON SO SCREEN MAY BE EDITED AGAIN
;BUT IF THE ESCAPE KEY IS PRESSED, IT RETURNS TO DOS
NEW:    MOV      AX,0
        INT      16H
        CMP      AL,27
        JZ       EXIT
        DEC      COMPLETE
;TURN CURSOR BACK ON
        MOV      AH,01    ;SET CURSOR TYPE REQUEST
        MOV      CX,0B0CH ;REGULAR CURSOR
        INT      10H      ;LET BIOS SWITCH CURSOR ON
        JMP      CENTER
;----------------------------------------------------------------
;END OF PROGRAM--RESTORE CURSOR AND EXIT TO DOS
EXIT:   MOV      AH,01    ;RESTORE CURSOR
        MOV      CX,0B0CH ;REGULAR CURSOR
        INT      10H      ;LET BIOS SWITCH CURSOR ON
        RET
START   ENDP
CODE    ENDS
END     START
```

BIBLIOGRAPHY

Berlekamp, Elwyn, Conway, John, and Guy, Richard. *Winning Ways*, vol. 2. New York: Academic Press, 1982. Chapter 25 is Conway's description of Life and self-reproducing Life patterns.

Blair, C. "Passing of a Great Mind." *Life*, February 25, 1957. *Life*'s obituary of Von Neumann.

Brillouin, Léon. *Science and Information Theory*. New York: Academic Press, 1956. Contains an analysis of Maxwell's demon with historical survey.

Burks, Arthur W. (ed.). *Essays on Cellular Automata*. Urbana and Chicago: University of Illinois Press, 1970. Contains detailed descriptions of Von Neumann's self-reproducing automatons and two of Ulam's papers on recursively defined geometric objects. Photos show a three-dimensional "coral reef" made of wooden cubes at Los Alamos.

Campbell, Jeremy. *Grammatical Man: Information Theory, Entropy, Language, and Life*. New York: Simon & Schuster, 1982. Information theory in science. The author reports that Ulam's recursive games were inspired by Noam Chomsky's theory of generative grammar.

Davies, P.C.W. *The Accidental Universe*. Cambridge, England: Cambridge University Press, 1982. A concise, detailed survey of how large-scale phenomena depend on the values of the physical constants: pro-anthropic principle, but with arguments from both sides.

Dyson, Freeman J. "Time without end: Physics and biology in an open universe." *Reviews of Modern Physics*, vol. 51, no. 3, July 1979. One of the first papers to consider the long-term cosmological future.

243

Dyson assumes that protons do not decay and speculates about the future up to the year $10^{10^{76}}$.

"The Game of Life." *Time,* January 21, 1974. Somewhat irritable account of hackers vs. "precious" computer time in early 1970s. "Martin Gardner tells of one computer specialist who has a special panic button under his desk: whenever a supervisor comes into the room, the specialist can wipe the display screen clean . . ."

Gamow, George. *One Two Three . . . Infinity: Facts and Speculations of Science.* New York: Viking Press, 1947. A popular science book containing a description of his universal printing press.

Gardner, Martin. *Wheels, Life and Other Mathematical Amusements.* New York and San Francisco: W. H. Freeman, 1983. A collection of Gardner's "Mathematical Games" columns from *Scientific American,* including all the material on Life. The $3n+1$ problem, Conway's game of Hackenbush, and Ulam's map folding problem are covered in the same volume.

Heims, Steve J. *John Von Neumann and Norbert Weiner: From Mathematics to the Technologies of Life and Death.* Cambridge (Mass.) and London: M.I.T. Press, 1980. A dual biography; the most complete account of Von Neumann's life to date, drawing on recollections of associates. Herman Goldstine recalls that he and Von Neumann "used to listen at meetings to serious presentations, and we would count the number of times in a sentence or paragraph that the speaker said something which we would deem to be 'nebech.' Then we would flash the number at each other by holding up fingers."

Hofstadter, Douglas R. *Gödel, Escher, Bach: An Eternal Golden Braid.* New York: Basic Books, 1979. Gödel's incompleteness theorem and artificial intelligence. One of many amusing digressions looks at what happens when a television camera is turned on its own monitor.

Misner, Charles W., Thorne, Kip S., and Wheeler, John Archibald. *Gravitation.* San Francisco: W. H. Freeman, 1973. A huge book on general relativity, mostly tangential to the current discussion, containing Wheeler's musings on a logical basis for physical law.

Moore, Edward F. "Mathematics in the Biological Sciences." *Scientific American,* vol. 211, no. 3, September 1964. Overview of trivial and nontrivial self-reproduction. A photograph shows an experiment with Penrose's tiles.

Trefil, James S. *The Moment of Creation.* New York: Charles Scribner's Sons, 1983. Grand unified theories and big bang cosmology.

Ulam, S.M. *Adventures of a Mathematician.* New York: Charles Scribner's Sons, 1976. Ulam's autobiography, containing much material on Von Neumann.

Von Neumann, John. *Theory of Self-Reproducing Automata.* Urbana and Chicago: University of Illinois Press, 1966. Von Neumann's proof that patterns of 29-state cells can reproduce.

Wainwright, Robert T. (ed.). *Lifeline: A Quarterly Newsletter for Enthusiasts of John Conway's Game of Life,* nos. 1–11, 1971–1973. The source of most of the Life objects in this book. Issues may be obtained by writing to *Lifeline,* 12 Longvue Avenue, New Rochelle, N.Y. 10804.

Weinberg, Steven. *The First Three Minutes.* New York: Basic Books, 1977. How the microwave background radiation and an assumption of thermodynamic equilibrium led to the details of the early universe.

INDEX

Acorn, 104
Aesthetics, 98, 99
Aircraft carrier, 38
AND gate, 187, 201–4, 209
Antibodies, 107–8
Antimatter, 152
April, Robert, 105
Ashby, Ross, 14

Bakery, 167
Banks, Roger, 49
Barber pole, 44
Barge, 38
BASIC, 233, 235–37
Beacon, 40–41, 173
Beehive, 28, 34, 40, 46
  frequency of, 172
  shuttle and, 88
Beeler, Michael, 105
Belly spark, 81
Berlekamp, Elwyn, 196
B heptomino, 88–89, 165, 173
  unlimited growth and, 108, 111
Biboat, 173
Big bang, 142–43, 163
Big S, 173
Billiard-table oscillators, 44
Biloaf, 167
Biology. *See also* Life (living organisms)
  as derived from chemistry and
    physics, 17–18
  self-reproduction in, 17–18, 189–90
Bipond, 173
Black-body radiation, 72
Black holes, 161–63

Blinker, 27, 38
  frequency of, 172, 173
Block, 27–28, 38, 40, 198
  as external memory register, 209–13
  frequency of, 171, 173
Block and glider, 167
Blueprints, 131–32, 186–89, 218–21
Boat, 36, 173
Boltzmann, Ludwig, 56–60, 62, 97, 98,
    100, 101
Boomerang (spaceship phase), 81
Bosons, 145, 150
Breeder, 119–21
Brillouin, Léon, 63, 73, 76, 77
Broths. *See* Random fields
Brownian motion, Maxwell's demon
    and, 65
Burks, Arthur W., 182, 194

Carnot, Sadi, 54
Cellular model, Von Neumann's,
    185–87
Cellular states, self-reproducing
    machines and, 15–16
Century, 167, 173
Chance, 21
  entropy, information and,
    100–101
Choice in creation, 18–22
Christina, queen of Sweden, 17
Clausius, Rudolf, 53–55
Clock, 40–41, 173
  3–4, 140
Clock II, 44
Collins, George D., Jr., 234

Collisions
  of galaxies, 157–58
  of Life objects, 48, 81–85, 210–17
  of subatomic particles, 84–85
Complexity, cosmic, 22–23, 31, 62,
      231–32
  origin of, 142, 151
Complexity barrier, 194
Computers, 98
  information-theory definition of life
      and, 193–94
  Life (game) for, 30–32, 233–42
  pattern games for, 14, 133–41
  universal. See Universal computer
Conway, John Horton, 24–35, 46, 50,
      79–80, 85–86, 88, 103, 104, 110,
      119, 126, 127, 141, 210, 213, 218,
      222–24, 226, 227, 229. See also
      Life (game)
  on self-reproducing Life patterns,
      196–98
Copenhagen interpretation, 85
Coral reef game, 133–36
Corderman, Charles, 104
Cosmic history (or evolution), 145–63.
      See also Galaxy formation
  first split second of, 146–55
  future events in, 155–63
  GUT era in, 150
  GUT freezing in, 150–53
  inflationary model of the early
      universe, 151–52
  planets in, 154–55
  radiation and, 152–53
  slowing down of, 146–47
  stars in, 153–54
  unification of subatomic forces in,
      145–46
Cosmic time scale, 146–55
Cosmology, 143, 146. See also Cosmic
      history (or evolution)
  exponential notation used in, 148
Creation, 231
  Boltzmann's view of, 100
  choice in, 18–22
Crick, Francis, 18
Crystals, 192

Dead bodies as living matter, 192
Defects, GUT freezing, 151
Density fluctuations, galaxy formation
      and, 143
Descartes, René, 17, 125
Deuterium, 152
Dicus, Duane A., 159, 161
Disorder
  definition of, 58–59
  entropy as, 56–61

DNA, 18, 189–90, 192, 193
DNA polymerase, 190
Dyson, Freeman J., 146, 158

Eater, 38–40
  self-reproducing Life patterns and,
      196
Ecologist (Life pattern), 113
Einstein, Albert, 18, 19, 85, 144
Eisenhower, Dwight D., 181
Electromagnetic force, 144–45, 152
Electrons, 145, 159, 161
Electroweak force, 144–45, 150, 152
Empirical knowledge, limits of, 74–77
Energy, unification, 145
Engines (puffer train)
  B heptomino, 111
  flying machine, 115–16
Entropy, 55–61
  chance and, 100–101
  decreases in, 100
  as disorder, 56–61
  information and, 97–98
  information theory and, 67–74
Enzymes, 190
Equilibrium, thermodynamic, 60, 61
Evolution
  cosmic. See Cosmic history (or
      evolution)
  of self-reproducing Life patterns,
      222–26

Fermat's last theorem, 197
Fermions, 145, 150
Figure 8, 42
First law of thermodynamics, 54
  Maxwell's demon and, 63–64
Fleet, 167
Flotillas, 86–87, 216–18, 220, 221
Flying machine, 115–16
Forces, subatomic. See Subatomic
      forces
Fredkin's game, 136–37
Future of the universe, 155–63

Gabor, Dennis, 63
Galaxy collisions, 157–58
Galaxy formation, 142–43, 151–55,
      232
  density fluctuations and, 143, 153
  GUT freezing defects and, 151
  neutrinos and, 153
  turbulence and, 143
Galaxy (Life pattern), 44
Game theory, 178–79
Gamow, George, 92
Garden-of-Eden patterns, 49
Gardner, Martin, 24, 103

Glider, 30–31, 48, 78
  acorn and, 104
  antibodies and, 107–8
  asymmetry of, 79
  collisions of, 81–85, 198–201, 210–17
  frequency of, 172, 173
  R pentomino and, 33, 35
  self-reproducing Life patterns and,
    213–17
  shuttling, 44–45
Glider gun, 51, 105–8. *See also* Thin
    gun
  self-reproducing Life patterns and,
    196
Glider streams, 198–201
  AND, OR, and NOT gates and,
    201–4
  duplicating machine for, 204–8
Glider train, 119–21
God, 18, 19, 22
Gosper, R. William, Jr., 105, 106, 108,
    119, 196
Grand unified theories (GUTs), 20, 22,
    23, 126, 127, 161. *See also* GUT
    era; GUT freezing
  proton decay and, 158–61
  strong, electromagnetic, and weak
    forces in, 144–45
Gravity, 144, 145
  cosmic evolution and, 148, 150
Growth
  accelerating patterns of, 119
  unlimited. *See* Unlimited growth
GUT era, 150
GUT freezing, 150–53
Guy, Richard, 196

Hamlet (Shakespeare), 22–23
Hat, 173
Hawking, Stephen, 162
Heat. *See also* Thermodynamics
  theories of, 56
Heat death, 62
Heisenberg, Walter, 65, 78
Helium, 152
Heptominos
  B, 88–89, 108, 111, 165, 173
  Herschel, 34–35, 167, 170, 173
  Pi, 110, 165–66, 173
Herschel, 34–35, 167, 170
  frequency of, 173
Hexadecimal, 173
Hidden variables, 85
Higgs field, 151–52
High-entropy active regions, 171
Honey farm, 34, 46, 167, 212
Howell, Richard, 105
Hydrogen, 156, 158–60

IBM PC, Life (game) for, 237–38
Inflationary model of the early universe,
    151–52
Information
  chance and, 100–101
  maximum rate of transmission of, 97
Information demon, 90–97
Information theory, 22, 52, 126–27
  entropy and, 67–74, 97–98
  life as defined in, 190–94
  limitations of knowledge and, 76–77
  Shannon and, 90, 97–98
Interchange, 167
Irreversibility, 56–57

Jacobson, Homer, 130

Kickback reaction, 198–200, 206–7,
    215–16
Kinematic model of reproduction,
    182–85
Knowledge. *See also* Information
    theory
  limits of, 21–22, 74–77
  useful, 77

Laplace, Marquis Pierre Simon de,
    20–22, 61, 125–27
Letaw, John R., 159, 161
Life (game), 24–51, 126, 127
  accelerating growth patterns, 119
  beehive, 28, 34, 40, 46, 88, 172
  blinker, 27, 38, 172, 173
  block, 27–28, 38, 40, 171, 173, 198,
    209–13
  collisions, 48, 81–85, 198–201, 210–17
  for computers, 30–32, 233–42
  engineer approach to, 47–51
  on finite versus infinite fields, 35
  as forward-deterministic, 48–49
  frequencies of objects in, 171–75
  glider. *See* Gliders
  Herschel, 34–35, 167, 170, 173
  honey farm in, 34, 46, 167, 212
  naturalist approach to, 47–51
  oscillators, 40–46, 139
  pixel rows, 46–47
  played with checkers, 26
  played with graph paper, 26–27
  random fields. *See* Random fields
  reason for using, 31–32
  R pentomino. See *R* pentomino
  rules of, 25–27, 29–31, 78
  self-reproducing patterns. *See*
    Self-reproducing Life patterns
  shuttles, 88–89, 105–7
  spaceships. *See* Spaceships
  stabilization of fields, 174–76

Life (game) *(cont'd.)*
  still lifes, 36–40, 139, 173
  symmetry, 46–47
  3–4 Life, 138–41
  traffic light, 29, 34, 165, 166,
    172–73
  T tetromino. See *T* tetromino
  universal computers, patterns that
    function as, 197–217
  unlimited growth. *See* Unlimited
    growth
  unpredictability of patterns, 29,
    49–51
Life (living organisms), 156. *See also*
  Biology
  information-theory definition of life,
    190–94
  self-reproduction, 17–18, 189–90
Life force, 17, 18
Light, Maxwell's demon and, 72–73
Loaf, 36, 40, 173
Logic, 19
Long barge, 38
Lumps of muck, 167, 173

Macrostates, 59–60
Manhattan Project, 14
Mathematics, recursion in, 123–24
Maxwell, James Clerk, 52, 144
Maxwell's demon, 52–53, 61,
    100–101
  Brownian motion and, 65
  critiques of, 62–66
  energy required by, 64–65
  first and second laws of
    thermodynamics and, 53–55
  first law of thermodynamics and,
    63–64
  as information demon, 90–97
  light and, 72–73
  quantum uncertainty and, 65–68
  second law of thermodynamics and,
    62
Meaning, structure and, 98–100
Memory registers, 208–13, 218,
    220–22
Microstates, 59–60
Minsky, Marvin, 209
Morgenstern, Oskar, 179
Mule, 192

Natural laws, 77
Natural Life objects, 47–51
Neutrinos, 153, 159
Neutrons, 152
Newton, Sir Isaac, 144
Noah's ark, 118
NOT gate, 187, 202–4, 208

Observation
  entropy and, 67–74
  information theory and, 22
Order
  cosmic, chance and, 100–101
  definition of, 58–59
OR gate, 187, 202–4, 209
Oscillators, 40–46
  in 3–4 Life, 139

Paik, Nam June, 92
Paper clip, 173
Penrose, L.S., 128, 130
Penrose's tiles, 191, 192
Pentadecathlon, 43–46
Perpetual-motion machines, 52, 54. *See
    also* Maxwell's demon
  glider guns as, 105–6
Photons, 145, 152–53, 159
Physics
  comprehensive theory of nature,
    20–21
  as derivative, 20–21
  recursion in, 124–27, 231
  reductionist tradition and,
    20–21
Pi, 230–31
Pi heptomino, 110, 173
  in random fields, 165–66
Pixel rows, 46–47
Planck time, 148, 150
Planetary motions, 125–26
Planets
  future of, 157–60
  origin of, 154–55
  proton decay and, 159–60
Pluto, 160
Point defects, 151
Pond, 85, 106, 173
Positrons, 159, 161
Protons, 152
  decay of, 158–61
Puffer trains, 51, 110–18
  flying machine, 115–16
  switch engine, 116–18
Pulsar, 42, 173
Pulsators, 42–44

Quantum physics, 85
Quantum uncertainty, 21. *See also*
    Uncertainty principle
  Maxwell's demon and, 65–68
Quarks, 145, 152

Radiation
  black-hole, 161, 162
  cosmic evolution and, 152–53
  microwave background, 160

Random fields, 47–48, 50, 164–76
common unstable objects in, 165–71
density of, 164–65
final density of, 175
frequencies of Life objects in,
171–75
high-entropy active regions in, 171
self-reproducing Life patterns and,
222, 223
stabilization of, 174–76
Recursion, 122–32
in mathematics, 123–24
in physics, 124–27, 231
self-reproduction and, 127–31
Recursive games, 133–41
Fredkin's game, 136–37
Ulam's coral reefs, 133–36
Reductionism, 20, 21, 23, 31, 60
biology and, 17–18
Redundancy, 98
Ribosomes, 189–90
RNA, 189–90, 192
R pentomino, 33–36, 127, 137, 167,
170, 173, 176
growth of, 35
stabilization of, 35–36
unlimited growth and, 103–4

Salam, Abdus, 144
Schroeppel, Rich, 105
Second law of thermodynamics, 55, 57,
60, 62, 100. *See also* Entropy
Self-reproducing Life patterns, 24,
195–229
as alive, 226
AND, OR, and NOT gates and,
201–4
blueprints for, 218–21
evolution of, 222–26
glider stream duplicating machine
and, 204–8
glider streams and, 198–201
kickback reaction and, 198–200
memory registers and, 208–13, 218,
220–22
size of, 218, 226–29
supervisory unit of, 220–21
thin gun and, 199–201
universal computers, patterns that
function as, 197–217
universal constructor, 213–17
Self-reproducing machines, 14–16,
182–89
cellular model for, 185–87
cellular states and, 15–16
central problem of, 187–90
kinematic model of reproduction
and, 182–85

living cells as, 17–18
supervisory unit and, 188–89
trivial self-reproduction of, 128–30
universal Turing machine and,
183–84
Self-reproduction
central problem of, 187–90
complexity barrier and, 194–95
of Life objects. *See* Self-reproducing
Life patterns
nontrivial, 128, 130–31
paradox of, 131–32
recursion and, 127–31
trivial, 128–30
universal constructor and, 130–32,
186–93
Shakespeare, William, 23
Shanks, William, 124
Shannon, Claude, 14, 67, 90, 97–98
Ship, 36, 173
Shiptie, 167
Shuttles, 88–89
unlimited growth and, 105–7
Side-tracking, 214–17
Silk, Joseph, 153
Simulation games, 127
Slater, J.C., 65
Smith, Alvy Ray, III, 49
Smoluchowski, M. von, 65
Solar system, 154–55
Soviet Union, 180–81
Space rake, 113–15
Spaceship factories, 108
Spaceships, 78–89. *See also* Glider
asymmetry of, 79–80
boomerang and dense phases of, 81
collisions with, 81–85
flotillas of, 86–87
frequency of, 172, 173
lightweight, middleweight and
heavyweight, 80–81, 108
overweight, 85–87
sparks thrown off by, 81
speed of, 79–80
in 3–4 Life, 140
Sparks, 81
Speciner, Michael, 105
Stairstep hexomino, 141, 167, 173
Stars
birth and death of, 156
formation of, 153–54
future of, 156–58, 160
life-spans of, 154
Still lifes, 36–40
frequency of, 173
in 3–4 Life, 139
String defects, 151
Strong force, 144–45, 150

Structure
  information and, 90
  meaning and, 98–100
Subatomic forces, 143–46, 152
  during GUT era, 150
  unification of, 144–46
Subatomic particles
  collisions of, 84–85
  hidden variables of, 85
Sub-quantum particles, Maxwell's
    demon and observation with,
    73–74
Super-grand unified theory, 20
Super-grand unified theory (SGUT) era,
    148, 150
Supernovas, 154
Supervisory unit, 188–89, 195
  of self-reproducing Life patterns,
    220–21
Switch engine, 116–18
Symmetry, in Life (game), 46–47
Szilard, Leo, 21–22, 67–74, 76, 77

Tail spark, 81
Teplitz, Doris C., 159, 161
Teplitz, Vigdor L., 159, 161
Thatcher, J.W., 194
Thermodynamic equilibrium, 60, 61
Thermodynamics, 61–62
  Clausius's studies in, 53–55
  first law of. See First law of
    thermodynamics
  information and, 97–98
  second law of. See Second law of
    thermodynamics
Thin gun, 199–201
  glider stream duplicating machine
    and, 205–8
Thompson, Hugh, 171
$3N + 1$ problem (Ulam's problem), 122
3–4 Life, 138–41
Time machine, 147–48
Time scale, cosmic, 146–55
Toad, 40–41, 173
Toroidal displays, 234
Traffic light, 29, 34, 165, 166
  frequency of, 172–73
Truman, Harry S, 181
T tetromino, 28–30, 46, 173
  flying machine and, 115–16
  pentadecathlon derived from, 43
  in random fields, 165
Tub, 38, 46, 173
Tumbler, 44
Turbulence, origin of galaxies and, 143
Turing, Alan, 180, 183
Turing machine. See Universal
    computer (universal Turing
    machine)

Ulam, Stanislaw M., 14–15, 25, 122,
    126, 180, 185
  coral reef game, 133–36
Ulam's problem ($3N + 1$ problem),
    122
Uncertainty principle, 85. See also
    Quantum uncertainty
  black holes and, 161, 162
Unification energy, 145
Unified theories. See Grand unified
    theories (GUTs)
Universal computer (universal Turing
    machine), 183–84, 187, 195
  Life patterns that function as,
    197–217. See also
    Self-reproducing Life patterns
  memory registers of, 209
Universal constructor, 130–32,
    186–95
  Life, 213–17
  in living systems, 189–93
Unlimited growth, 103–21
  acorn, 104
  breeder, 119–21
  glider guns, 105–8
  puffer trains, 51, 110–18
  strategies for, 103–5
  switch engine and, 116–18
Unpredictability of Life patterns, 29,
    49–51

Viruses, 192
Von Neumann, John, 13–18, 24, 67,
    85, 97–98, 124, 126, 128,
    177–96, 196, 209, 226–27,
    230–32
  biographical sketch of, 177–82
  cellular model of, 185–87
  central problem of self-reproduction
    and, 187–90
  complexity barrier and, 194–95
  game theory, 178–79
  information-theory definition of life
    and, 190–94
  kinematic model of reproduction
    and, 182–85
  on trivial and nontrivial
    self-reproduction, 128–31

Walter, W. Grey, 14
Watson, James, 18
Weak force, 144–45, 152
Weather prediction, 124–25
Weinberg, Steven, 144
Wheeler, John A., 150

Zel'dovich, Yakov, 153
Zen for Film, 92, 95, 96